Supersonic Cruise Technology

NASA SP-472

Supersonic Cruise Technology

F. Edward McLean

NASA Scientific and Technical Information Branch   1985
National Aeronautics and Space Administration
Washington, DC

Library of Congress Cataloging in Publication Data

McLean, F. Edward.
    Supersonic cruise technology.
    (NASA SP ; 472)
    Bibliography
    1. High-speed aeronautics — Research — United States.
I. Title. II. Series.
TL551.5.M35 1985   629.132'305   83-26912

For sale by the Superintendent of Documents, U.S. Government Printing Office
Washington, D.C. 20402

Foreword

The National Aeronautics and Space Administration and its predecessor, the National Advisory Committee for Aeronautics (NACA), have participated since 1920 in efforts to develop the technology required for supersonic cruise flight. Preliminary work concentrated on developing rudimentary test facilities and methods that would permit the investigation of supersonic problems. This was accompanied by research for defining aircraft and propulsion concepts for flight at speeds greater than the speed of sound. These early investigations contributed to the development of the joint U.S. Air Force/Navy/Bell XS-1 airplane that was piloted on the first successful supersonic flight by Air Force Capt. Charles E. "Chuck" Yeager in 1947.

Between 1956 and 1971, a strong research effort supported the USAF supersonic B-70 and commercial supersonic transport concepts. After neither of these programs resulted in a production aircraft because of technical and political problems, NASA was given the responsibility of establishing the technology base for a viable supersonic cruise airplane. This latter effort, known as the NASA Supersonic Cruise Research (SCR) program, was conducted from 1971 to 1981. The NASA Variable Cycle Engine (VCE) program, a propulsion offshoot of SCR, was conducted from 1976 to 1981.

The SCR program was somewhat unusual for a NASA program involving in-house and contractor participation. Several of the manufacturers provided company manpower and dollars to augment NASA funding and personnel, even though there was no contractual requirement for them to do so. Dissemination of technical information was almost immediate to all participants because data dumps and separate status reports were mutually transferred openly and with high frequency, and because of the day-to-day contacts by the major civilian and military airframe and propulsion system manufacturers. The disciplinary research/systems integration approach of the SCR and VCE programs led to a large number of major advances in the technology needed for supersonic flight.

This document provides a historic perspective of supersonic cruise technology, beginning with the early NACA supersonic research and including efforts during the B-70 and SST phase. It also records technological progress made in the NASA SCR and VCE programs. Since every research result could not be detailed here, only the most critical technology issues and research findings are presented.

WILLIAM S. AIKEN, JR.
Director, Aeronautical Systems.
Division
Office of Aeronautics and Space
Technology

Acknowledgments

The author wishes to express his appreciation to the following people for their assistance in the preparation of this document: Cornelius Driver, Domenic Maglieri, Sherwood Hoffman, Betty Pate, and Donald Hearth of the NASA Langley Research Center; Frank Neumann and Wilbur Middleton of the Boeing Airplane Company; Richard FitzSimmons and Bill Rowe of the Douglas Aircraft Company; John F. McCarthy, Jr., of the NASA Lewis Research Center; Clarence A. Syvertson of the NASA Ames Research Center; Isaac T. Gillam IV of the NASA Dryden Flight Research Facility; and William Pomeroy of NASA Headquarters. The author also expresses his appreciation to Donald D. Baals of the NASA Langley Research Center for his excellent unpublished history of NACA/NASA participation in the early supersonic technology efforts leading up to the SST effort.

Contents

SUPERSONIC CRUISE TECHNOLOGY

CHAPTER 1

Introduction

The National Advisory Committee for Aeronautics (NACA) was established by the United States Congress in 1915 as the aeronautical research arm of the U.S. Government. NACA's principal goal was to establish and maintain a preeminence for America in the field of aeronautics, an assignment that NACA performed with distinction until 1958. At that time a new agency, the National Aeronautics and Space Administration (NASA), was formed by Congress to absorb and continue the aeronautical duties of NACA and to take on the additional responsibility of conducting a program for the exploration of outer space. One of NASA's chartered goals is "the preservation of the role of the United States as a leader in aeronautical and space science and technology and in the application thereof to the conduct of peaceful activities within and outside the atmosphere." [1] *

For more than 60 years, first NACA and then NASA has endeavored to live up to its assigned goals. Working independently or in concert with the military services and other organizations, scientists and technicians at NACA/NASA have solved or assisted in the solutions of the most complex problems associated with flight. The solutions of these problems have led to dramatic extensions of the boundaries of manned and unmanned flight in the atmosphere and in outer space. In addition, while solving the problems, NACA/NASA has

---

* Superscript numbers refer to references, which are listed at the end of this document.

developed and acquired unique experimental facilities for use in the investigation of almost every facet of flight.

A substantial portion of the aeronautical research effort at NACA/NASA has been devoted to the consideration of problems associated with manned flight at supersonic speeds up to four times the speed of sound. The purpose of this research was to develop a technology base that would permit the military services and the aerospace industry of the United States to take full advantage of the recognized potentials of high-speed flight — more rapid response, improved offensive capability, increased survivability for military aircraft, reduced trip times, improved comfort, and increased productivity for commercial air transports.

The NACA/NASA program to provide a continuously viable supersonic technology base has evolved in three fairly distinct phases. First, there was a preliminary supersonic technology effort that began in the mid-1930s during the NACA era and lasted almost until NACA was

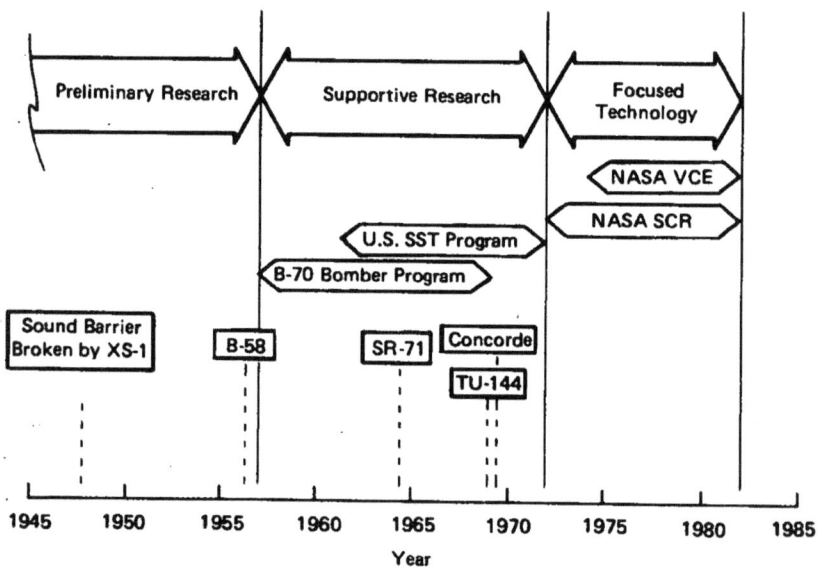

Phases of NACA/NASA research on supersonic cruise technology.

INTRODUCTION

absorbed by NASA in 1958. In this period, NACA developed experimental methods and facilities for the study of supersonic problems and contributed to the technology base used to prove the feasibility of manned supersonic flight. Later, this technology was part of the data base used in developing a number of military airplanes such as the century-series fighters, the B-58 supersonic dash bomber, and the SR-71 supersonic reconnaissance airplane, which is still operational. Although these airplanes demonstrated many of the potential advantages of supersonic flight, they all had limited payload capability or relatively short nonstop flight distances unless refueled in the air.

After the feasibility of supersonic flight was established, NASA raised its research sights to consider the difficult technical problems posed by the concept of supersonic cruise flight (i.e., sustained, unrefueled supersonic flight of a large-payload airplane over transcontinental and intercontinental distances). This second phase of the NACA/NASA supersonic technology effort began shortly before the formation of NASA in 1958 and lasted until 1971. During this period, NASA's principal task in supersonic technology was to conduct specific research in support of America's efforts to develop a supersonic cruise bomber (the B-70) and a commercial supersonic transport (the U.S. SST). Initiated in 1957, the B-70 was essentially canceled 2 years later when the decision was made to depend on intercontinental ballistic missiles (ICBMs) for America's strategic defense. The program for developing a U.S. SST began in 1963 and ended in 1971 in a sea of controversy involving a mélange of technical, environmental, political, and social issues.

Many people assumed that the cancellation of the U.S. B-70 and SST programs, along with the limited success of foreign SST efforts, provided positive proof that the promised advantages of supersonic cruise flight do not exist. Others believed that the advantages existed but that ways for using them had not been found. This latter belief led to the third phase of the NACA/NASA supersonic technology effort—a focused research program for solving the remaining technical problems that have inhibited the general acceptability and applicability of the concept of supersonic cruise flight. This program, which became

3

known as the NASA Supersonic Cruise Research (SCR) program with a supersonic propulsion offshoot called the NASA Variable Cycle Engine (VCE) program, spanned the period between 1971 and 1981. During this period, NASA conducted a broad attack on the problems of supersonic cruise flight with the support of the U.S. aerospace industry.

In each of three phases of supersonic cruise technology effort, NACA/NASA made vital contributions to the supersonic "state of the art." The early NACA preliminary research contributed substantially to the present understanding of high-speed flight and helped to open the door to the Moon and the planets. Although not assigned a decision-making role in America's aborted attempts to develop the B-70 and SST, NACA/NASA provided supporting research and concepts that were very important to these programs. More recently, the NASA SCR and VCE programs have made outstanding progress in identifying solutions to the major problems associated with supersonic cruise flight. In fact, these latter programs have brought the technology base for an environmentally acceptable and economically competitive supersonic cruise transport very near at hand.

Notwithstanding these important contributions, the NACA/NASA supersonic cruise technology program has not measured up to the success normally expected of a NACA or NASA program. There is currently no large-payload, long-range supersonic cruise aircraft in the air with an American label, and there seems to be little prospect for one in the future. With the cessation of the NASA SCR and VCE programs in 1981, no focused United States efforts remain for providing the technology for supersonic cruise aircraft. In spite of NACA and NASA's outstanding contributions to the understanding of supersonic cruise flight, the future of this concept is uncertain at best.

This rather bleak prognosis for the concept of supersonic cruise flight has tended to dim the lustre of early NACA research in this area. The spirited controversies that swirled around the B-70 bomber and U.S. SST programs, and eventually led to their cancellation, have obscured the NACA/NASA contributions to supersonic technology during these two programs. And the continuing SST controversy has perhaps

INTRODUCTION

depreciated the outstanding progress in supersonic cruise technology that has resulted from the recent NASA SCR and VCE programs.

The purposes of this document are to provide a brief perspective of the early NACA supersonic research, to review the NACA/NASA supersonic technology efforts during the B-70 and SST programs, to discuss the factors and events that led to the formulation of the NASA SCR and VCE programs, and to record the supersonic cruise technology progress that has been made in these latter programs. No attempt is made to provide a detailed history of supersonic cruise technology or to single out every research result. Rather, consideration is given to the most critical technology issues and research findings.

The character of NASA's research on supersonic cruise flight has been influenced by the arguments for and against the concept, and by the ebb and flow of the nation's commitment to supersonic cruise development. Chapter 2 considers these arguments and commitments and essentially "sets the stage" for supersonic cruise technology efforts. The remaining chapters examine the results and ramifications of these technology efforts.

Setting the Stage for Supersonic Cruise Technology

After his first successful flight in a powered heavier-than-air vehicle in 1903, man continued to search for ways to fly faster and faster. This quest for speed was motivated by several factors. The early daredevil pilots sought, perhaps, a few days in the record book, the cheers of the crowd, or the satisfaction of flying faster than anyone else. The military services wanted faster airplanes to gain a tactical edge over the real and potential enemies in the sky. The airline operators looked for faster, larger transport airplanes to attract more passengers and to satisfy steadily increasing passenger demand. Test pilots and aeronautical scientists were driven to investigate the unknowns of faster speeds to extend the usable boundaries of flight for both civilian and military airplanes. Spurred by these and other motivating forces, the quest for speed led to more sophisticated airplanes, to more powerful engines, and, by 1946, to flight speeds that approached the mythical speed barrier—Mach 1, the speed of sound.

THE SOUND BARRIER

By 1946, there was general, but not universal, public opinion that the speed of sound in air—760 miles per hour at sea level and 660 miles per hour at an altitude of 36 000 feet—represented an impenetrable barrier through which no airplane could fly.[2] A number of airplanes and pilots had been lost during flights at velocities near the speed of sound, and others had experienced strong buffeting and loss of control.

7

SUPERSONIC CRUISE TECHNOLOGY

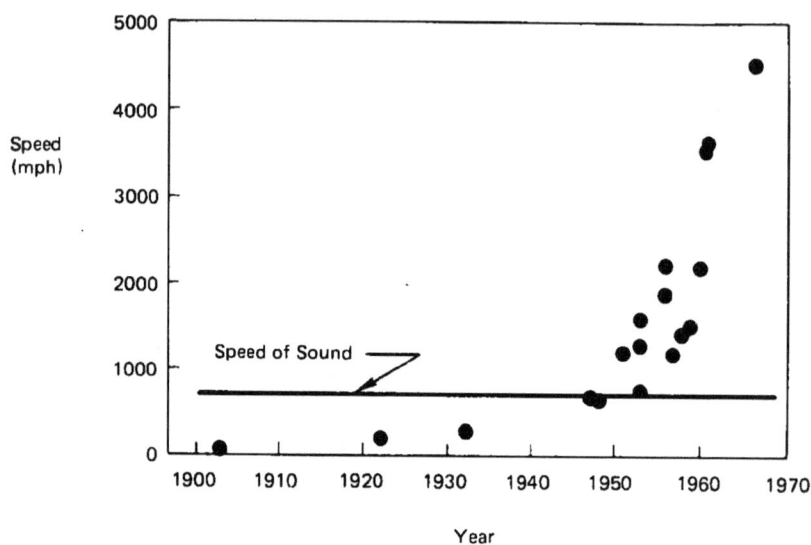

Year	Pilot	Speed (mph)	Year	Pilot	Speed (mph)
1903	Orville Wright	7	1956	Frank Everest, Jr.	1900
1922	William Mitchell	223	1956	Milburn Apt	2250
1932	James Doolittle	294	1957	Adrian Drew	1208
1947	Charles Yeager	700	1958	Walter Irwin	1404
1948	Richard Johnson	671	1959	Joseph Rogers	1526
1951	William Bridgeman	1238	1960	Joseph Walker	2196
1953	Frank Everest, Jr.	755	1961	Robert White	3603
1953	Scott Crossfield	1291	1961	Joseph Walker	3645
1953	Charles Yeager	1600	1967	Peter Knight	4564

America's quest for speed.

SETTING THE STAGE

Great Britain had abandoned all efforts to develop manned research airplanes for flying through the sound barrier, citing as a reason, among others, that "We have not the heart to ask pilots to fly the high-speed models, so we will make them radio-controlled." [3] In the United States, however, a research effort was under way on the rocket-powered Bell X-1, which would lead to the first successful flight of a manned aircraft through the so-called "sound barrier" on October 14, 1947, by U.S. Air Force Capt. Charles Yeager.

The first flight at supersonic speeds was a pivotal event in aviation history. Man soon found that he could design and construct airplanes that passed smoothly through the once formidable "sound barrier" and onward to speeds that pushed him to the outer fringes of the atmosphere and into space. In the early 1950s, supersonic flight of military "century series" fighter planes became commonplace, and the U.S. Air Force developed the B-58 bomber, which had a supersonic "dash" capability of Mach 2 — twice the speed of sound. First flown in 1956, this airplane signaled the approach of the era of supersonic cruise flight.

The years leading to the first supersonic flight were exciting years in the progress of aviation technology in the United States. Technology advances in the speed and versatility of aircraft were supported not only by the government, but by wealthy patrons, organizations, and individual entrepreneurs lured by the prospects of worldwide fame and cash prizes offered in national and international competitions. As aircraft entered the supersonic speed regime, however, the escalating costs of further increases in speed appeared to place future high-speed technology strictly in the province of the U.S. Government.

THE FIRST SUPERSONIC CRUISE AIRCRAFT

The first successful flight at supersonic speeds, along with subsequent experience gained in fighters and the B-58 program, led the U.S. Air Force to consider the development of a large-payload, long-range supersonic cruise aircraft. On December 23, 1957, the North

9

First "real" supersonic cruise aircraft—the U.S. Air Force North American B-70 bomber.

U.S. Air Force Lockheed YF-12—companion to the SR-71 that had supersonic cruise capability but without appreciable payload.

SETTING THE STAGE

American Aviation Company (now Rockwell International) was placed under contract to design and produce the B-70 supersonic bomber, an airplane that was to have intercontinental range capability at a sustained supersonic cruise speed of Mach 3 — three times the speed of sound. At the announcement of the B-70 contract award, factories across the country began to hum in a massive project that would soon involve 8000 contractors spread throughout the nation.[4] However, this flurry of activity was to subside rather rapidly since, only 2 years after the B-70 contract was issued, the Eisenhower administration essentially canceled the program with the justification that "the age of manned bombers was doomed by intercontinental ballistic missiles."[5]

Many members of the U.S. Congress were not happy with the cancellation of the B-70 bomber program, and the battle between the Congress and the White House lasted through the remainder of the Eisenhower administration and into the administration of President John F. Kennedy. During attempts to keep the B-70 program alive, the Preparedness Investigating Subcommittee of the powerful Senate Committee on Armed Services recommended in a report dated July 8, 1960, that "unless an operational supersonic bomber is developed now there will be no replacement for the B-52 at the time at which it enters its period of obsolescence (mid-1960s)."[6] It was 22 years later, in 1982, that the Reagan administration selected the Rockwell B-1 bomber as a replacement for the B-52.

All efforts to restore production funding for the supersonic B-70 bomber were unsuccessful, and the program was limited to the construction of two prototype aircraft. In 1964, the U.S. Air Force announced the secret development of a high-altitude reconnaissance airplane, the SR-71, which had a supersonic cruise speed of Mach 3. Although this airplane is still in operation, production was limited, and the B-70 bomber program represented the last major interest of the U.S. military services in long-range supersonic cruise flight.

THE NEXT LOGICAL STEP

Although the July 1960 report of the Preparedness Investigating Subcommittee did not succeed in its primary goal of securing production funding for the B-70, the report left a ray of hope for the supersonic cruise aircraft concept with these words:

> It is generally believed by air transportation authorities that a Mach 3 cruise transport represents *the next logical step* [author's italics] beyond the present family of jet transports. For both economic and prestige reasons, other nations, including the Soviet Union, will be interested in taking this step ahead of the United States. The impact of a supersonic transport on world traveling habits and on the entire field of transportation is expected to be far greater than the impact that accompanied introduction of current jet transports. Our leadership in commercial aviation will almost certainly be lost unless the Nation continues the development of Mach 3 technology and applies it as promptly as possible to air transport adaptations. [7]

NASA had expressed a view similar to that of the subcommittee in a report dated June 1960, which stated: "The successful development of a supersonic transport is of vital importance to the national prestige as well as the commercial stature of the United States." [8] Six months later, the Federal Aviation Agency recommended "that the Executive and Legislative branches of our government give prompt and careful attention to the immediate establishment of a national program for the development of a commercial supersonic transport aircraft." [9] The pressures for a U.S. supersonic cruise airplane program appeared to be mounting, with or without the B-70.

Was the development of a commercial supersonic cruise transport *the next logical step* in the evolution of the air transportation system? It certainly appeared to be in light of the situation that existed in the early 1960s. Among the factors that tended to favor this step were:

- The introduction of American-built subsonic jet transports in 1958 and 1959 had revolutionized the world's air transportation industry. The added speed and comfort of these jet airplanes over the existing propeller-driven transports led to a rapid increase in airline passenger demand and much improved airline productivity. Due to the popularity of its new subsonic jet transports, the American aircraft industry captured a dominant share of the world transport market, and the export sales of subsonic jet transport aircraft became an increasingly important positive factor in the American foreign balance of trade.

- The development of a commercial supersonic transport (SST) for airline service in the early 1970s would be properly timed to replace the subsonic jet transports that were just going into service.

- Programs for developing a supersonic cruise transport had been announced by the Soviet Union (TU-144) and by a consortium of the British and French governments (Concorde). Both of these programs were to receive government funding support and, without a competing American aircraft, could threaten United States domination of the world's air transport market.

- Although the production program for the supersonic B-70 had been canceled, two of the advanced airplanes were to be constructed. It appeared probable that research results and design information obtained on these two large B-70 airplanes would be applicable to an SST and would therefore effectively reduce the development cost of such an aircraft.

- The American X-1 airplane was the first manned vehicle to fly at supersonic speeds. Since that flight, the military services and NASA had conducted many supersonic wind tunnel and flight tests. By 1963, manned American winged vehicles had flown to altitudes of 354 200 feet (67 miles) and at speeds of 4100 miles per hour (six times the speed of sound). These altitude and speed marks were recorded by an experimental aircraft, the X-15, and were not directly indicative of America's ability to develop an SST. However, the achievements of this experimental aircraft did indicate the advanced level of the nation's high-speed aircraft technology.

- The U.S. Congress had provided the military services and NASA with funding for acquiring unique research facilities that could be used in investigating nearly every aspect of supersonic cruise flight. In NASA alone, the research facilities for transonic and supersonic investigations represented an investment of over $180

13

million.[10, 11] By the early 1960s, these research facilities had been widely used in the supersonic B-70 bomber program and in the investigations of potential SST configurations.

- Both NASA and the American aircraft industry had a large number of experienced research engineers and scientists who had been involved in supersonic research and flight investigations.

These factors, along with others that were present in the early 1960s, indicated the readiness of the United States to pursue the development of an SST and take what appeared to be the next logical step in the evolution of air transportation. However, negative factors were also present at this time, and many questioned the wisdom of taking the step. Among these negative factors were:

- By 1963, America had placed manned spacecraft into Earth orbit and was well into a massive program for landing a manned spacecraft on the Moon by the end of the decade. The development of a commercial SST would also be a big federal undertaking, and it was questionable whether two programs of such technical magnitude could be successfully managed and funded at the same time. The two programs would also tend to compete for support and interest.

- Historically, military airplane development had served as a pacesetter for technical developments that were subsequently used by the commercial sector. Because the military services had not fully developed a supersonic cruise aircraft, any such venture would have to be justified by the commercial application alone. Without military participation, the cost of development for commercial application would be much higher.

- Even with military support, the cost of supersonic cruise aircraft development would be beyond the means of the U.S. airplane industry, and hence, government financial support would surely be required. Never before had the government provided direct financial support in the development of a commercial airplane.

- The airlines had just invested heavily in the new subsonic jet transports and were less than enthusiastic about another new airplane, even a decade in the future.

- A number of major technical problems, such as engine noise, sonic boom, off-design performance, and operating flexibility, were still to be solved before an SST would be economically viable and environmentally acceptable. Although it was generally believed that these problems could be solved, they represented a substantial technical risk.

These and other issues relating to the advisability of an American program for developing a commercial SST were debated both inside and outside the government during the early 1960s. The U.S. Congress provided funding for preliminary SST research in fiscal year 1961, and announcements of the Russian and British/French SST programs came in 1962 and 1963, respectively.

The announcement of an American SST program came on June 5, 1963, when the late President John F. Kennedy, speaking at Air Force Academy graduation exercises, told the assembled group:

> As a testament to our strong faith in the future of airpower, and the manned airplane, I'm announcing today that the United States will commit itself to an important new program in civilian aviation. . . . It is my judgment that the government should immediately commence a new program in partnership with private industry to develop at the earliest practical date the prototype of a commercially successful supersonic transport superior to that being built in any country in the world. . . . This commitment, I believe, is essential to a strong and forward-looking nation and indicates the future of the manned aircraft as we move into the missile age as well. . . . [12]

With this recommendation by President Kennedy and approval by the U.S. Congress, it appeared that the United States was finally ready to take *the next logical step* in the evolution of air transportation — the development of a commercial supersonic cruise transport. The Federal Aviation Agency (FAA), which had been assigned management responsibility for the new program, followed up quickly after receiving the go-ahead from Congress. The FAA presented a Development Plan on June 19, 1963,[13] and issued a "Request For Proposals" on August 15, 1963.

The FAA plan for commercial SST development called for two or three phases. Phase I was an initial design competition between all interested airframe and engine manufacturers. If a clearly superior "winning combination" resulted from this competition, these contractors would be awarded contracts for the development program by May 1, 1964. If there was not a clear winner, a phase II competition would be interjected between the two leading airframe and engine manufacturers from phase I. The final selection of the airframe/engine combination to go into the development phase would then be made early in 1965.[14] As a matter of fact, the phase II "winning combination" was not selected until December 31, 1966, almost 2 years behind schedule.

The winning design in the FAA SST competition was the Boeing Commercial Airplane Company concept, the B-2707-200. The proposed airplane was to cruise at a supersonic speed of Mach 2.7 (nearly 1800 miles per hour) and was estimated to weigh 675 000 pounds. The airplane made use of the then controversial variable-sweep or swing-wing concept now incorporated on the F-111 and F-14 airplanes. In this concept, the wing can be swept during flight from a forward position for low-speed flight conditions to a rear position for supersonic cruise flight. Boeing saw this complex arrangement as the best means of achieving low noise at takeoff and landing and good flight efficiencies at all parts of the flight spectrum. The B-2707-200 was to be powered by four General Electric Company GE-4 turbojet engines that used afterburners (i.e., additional fuel was burned in a duct at the rear of the engine to provide increased power at takeoff and during acceleration). On April 29, 1967, the FAA awarded contracts to Boeing and General Electric to proceed with prototype development of the B-2707-200 airplane.[15]

After working on the design problems involved in the complicated variable-wing-sweep mechanism for about 1 year, Boeing concluded that the concept could not be integrated into a viable supersonic transport within the contract weight limitations.[16] Consequently, in April 1968, Boeing asked for and received the FAA's permission to

16

Lockheed L-2000 concept in competition with Boeing in U.S. SST competition.

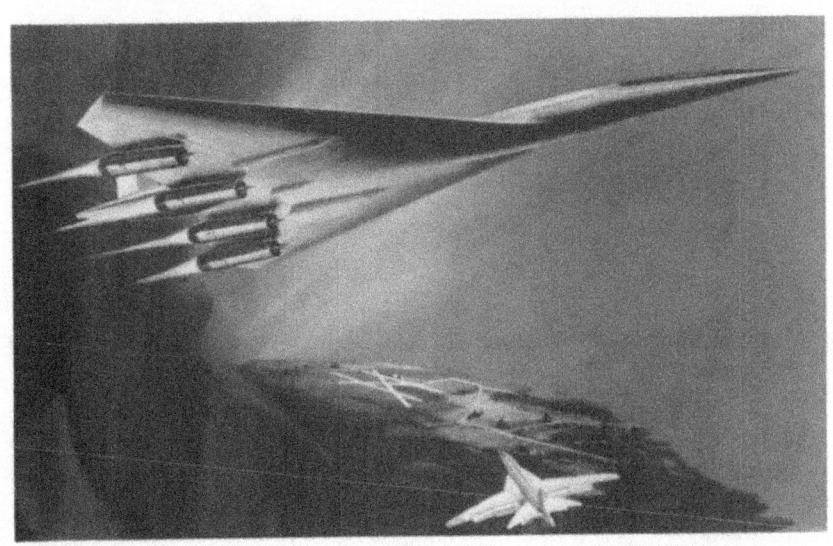

Boeing/General Electric 2707-200 concept that won the U.S. SST competition.

change to a simpler fixed-wing concept. This concept, a double-delta wing configuration labeled the B-2707-300 (Dash-300), was now to become the American entry in the SST sweepstakes. At the time of Boeing's decision to switch to the Dash-300, the other two entries in the SST race, the Soviet TU-144 and the British/French Concorde, were nearing flight status. The TU-144 made its maiden flight on December 31, 1968,[17] and the Concorde took to the air on March 2, 1969.[18]

The American program for developing an SST was a more demanding technical challenge than those for the TU-144 and the Concorde. The U.S. decision to leapfrog the foreign supersonic versions by going to a larger size and higher speed brought forth temperature, materials, and potential operating problems not experienced by the foreign designers. Delays by the FAA in selecting an airplane for development and by Boeing in generating an acceptable design were perhaps inevitable. These delays, along with increasing public concern over environmental issues, allowed opponents of the SST effort to mass forces. With each passing year, in spite of strong White House support, votes appropriating SST funding got closer and closer in the U.S. Congress. Finally, on March 24, 1971, the U.S. Senate voted 51 to 46 to cancel the SST development program.[19] On May 12, 1971, the House of Representatives made a last-ditch move to continue the program at least to the end of June 1971. The Senate refused to go along with the House, and the American SST program finally died on the night of May 19-20, 1971.[20]

AFTER THE U.S. SUPERSONIC TRANSPORT PROGRAM

By the time Congress terminated the SST program, over $1 billion had been spent on it, but the social costs of the termination were also high. Seven thousand Boeing workers and 6000 General Electric workers were laid off immediately. Seattle, Washington, with a 15 percent unemployment rate, became a mini disaster area.[21] There was also the question: What would happen to America's future interest in high-speed research and flight? The wide-body subsonic jet transports had been put into airline service in 1969 and would serve air transportation's productivity needs for the near term. But could an industry that had increased its productivity by 500 percent in the past 15 years while doubling its average speed continue to grow if the speed remained static? What would happen to the American airplane industry if improved, economically viable versions of the British/French Concorde and/or the Russian TU-144 were developed?

Following the cancellation of the American SST program, the Nixon administration and some members of Congress were concerned about these questions. The military services had shown no further interest in long-range supersonic cruise flight after the B-70 bomber program. Now, without the focus of the national SST program, there would likely be little emphasis on supersonic cruise research by the airframe manufacturers, the propulsion industry, or NASA.

Early in 1972, to fill this void in supersonic cruise research, the Nixon administration instructed NASA to formulate a focused supersonic technology program. The goal of this program was to provide answers to the environmental, performance, and cost problems of an SST and to establish a state of "technology readiness" within 3 to 4 years. The administration provided $11.7 million for this program in fiscal year 1973, and President Nixon requested $46 million in the fiscal year 1974 budget.[22] Opponents of the SST saw this program as a move to "restart the SST" and cut the funding back to $10.1 million for fiscal year 1974, but at least the program was under way.

The supersonic technology program that evolved from the concerns of the White House and some members of Congress was first called the NASA Advanced Supersonic Technology (AST) program and has subsequently been known as the NASA Supersonic Cruise Aircraft Research (SCAR) program and the NASA Supersonic Cruise Research (SCR) program. This program, hereafter referred to as the SCR program, operated continuously as a focused technology program in the NASA budget from fiscal year 1973 to fiscal year 1981 at a funding level of $8 million to $11.7 million per year. In fiscal year 1975, the SCR program identified a new propulsion concept, the "variable cycle engine," which showed promise of solving the noise and off-design operating problems of an SST. The variable cycle engine (VCE) study was broken out as a separate focused program under the management of the NASA Lewis Research Center. The VCE program operated from fiscal year 1976 to fiscal year 1981 at funding levels ranging from $0.5 million in the initial year to approximately $6.0 million in the peak years, FY 1979 and FY 1981.

As focused efforts within NASA, the SCR and VCE programs took a fresh look at all of the problems exposed in the U.S. and foreign SST programs—excessive noise, marginal range/payload capability, high cost, poor economics, etc. The major airframe manufacturers, Boeing, Douglas Aircraft Company, Lockheed-California Company, and Rockwell International, and the major propulsion system companies, General Electric and Pratt and Whitney Aircraft Company, were called into a partnership of research and technology. Some of these companies added their own funds to those of the government to stretch the research capability of the programs,[23] and the results of in-house company-funded research were published and reported along with those of the government-funded technology effort. This provided for an almost unprecedented instantaneous exchange of technical information between companies competing in both military and civilian projects and led to very little duplication of effort.

A recent report summarizes the NASA SCR and VCE program results as follows:

> Dramatic progress has been made in the major technical discipline areas, making it possible to show large gains in range/payload capability for supersonic transport type airplanes. At the same time, stringent environmental noise and pollution restraints can be met. Even if the supersonic airplanes should cost twice as much per pound as their subsonic competition, the supersonic vehicles may nearly equal subsonic tourist ticket prices while reducing trip time to less than half its present value. Two very recent developments, as yet unproven, one structural, one aerodynamic, could provide the performance improvement margin necessary to end the subsonic jet dominance of the long-haul over-water passenger market. Advanced SSTs could displace subsonic jets on these routes just like the subsonic jets displaced the propeller airplanes.[24]

SETTING THE STAGE

Unfortunately, this report goes on to say that "funding limitations have forced cancellation of the focused government/industry approach. . . ." [25]

Now that we have "set the stage" for the various phases of the NACA/NASA research and technology effort on supersonic cruise flight, let us turn back to the beginning and consider some of the details of this effort.

CHAPTER 3

NACA Preliminary Supersonic Cruise Technology (?–1956)

No "red letter" date marks NASA's entry into the field of supersonic cruise research. In the late 1930s and early 1940s, NACA engineers and scientists were conducting preliminary studies of airplane and engine combinations that might possibly fly at velocities greater than the speed of sound. However, these studies were limited by the uncertainties that existed in the aerodynamic characteristics of airplanes that were considered for such flight. At that time, wind tunnels were of the closed-throat type, and as the speed of the airflow approached that of sound, the tunnels "choked." Shock waves forming off the test model and its supports would reflect off the tunnel walls, inhibiting accurate measurements of flow characteristics and behavior around the model. This condition persisted from about Mach 0.7 to about Mach 1.3, the very area in which the scientists were interested, the transonic region between subsonic and supersonic flight.[26] Without some better research data in the transonic region, the accomplishment of transonic or supersonic cruise flight seemed hopeless.

With the entry of the United States into World War II in December 1941, the means for removing the uncertainties of transonic and supersonic flight became critically important. Greater high-speed capability was incorporated in each new American fighter and attack airplane, and these airplanes were getting into difficulties because of local regions of transonic and supersonic flow, even though their forward speeds were still subsonic. "Army Air Forces and Navy combat squadrons suddenly found that their aircraft faced a new and frighteningly mysterious danger aside from the enemy. An epidemic of tail failures during dives appeared in three production high-speed aircraft, the Republic P-47 Thunderbolt, the Curtis SB2C Helldiver, and the Bell P-39 Airacobra." [27] Although NACA was able to recommend suitable fixes for these three aircraft, the aerodynamic uncertainties of flight at velocities near the speed of sound remained. The turbojet engine, introduced in an experimental aircraft in Britain in 1941, promised further increases in speed if the aerodynamic uncertainties could be resolved. By 1943, ". . . the only remaining barrier to high-speed flight — including flight faster than the speed of sound — appeared to be a knowledge barrier, where aerodynamicists could not predict with certainty what occurred in the turbulent speeds of transonic flight. The immediate problem facing aerodynamicists was how to arrive at accurate research tools and methods to derive needed data and strip the mystery from compressibility and transonic aerodynamics." [28]

MOUNTING AN ATTACK ON THE SOUND BARRIER

In the early 1940s, with no immediate prospects for solving the wind tunnel blockage problems at velocities near the speed of sound, NACA engineers developed several stopgap methods for obtaining transonic aerodynamic data. One of the methods involved the dropping of weighted bodies from high altitudes. "The bodies would then attain velocities equal to or faster than that of sound. Radar and visual tracking from the ground could determine the speed and path of the falling body." [29] Another method, developed by Robert Gilruth of the NACA Langley Laboratory, involved placing small models in the local transonic and supersonic flow fields that were known to exist over the wing of a North American P-51 "Mustang" during high-speed dives. [30] These methods — together with the later NACA rocket-propelled model work under Gilruth — provided vital data on drag variation through the speed of sound. [31] However, there was rising sentiment that a research airplane was the best way to secure the transonic data necessary for breaking the "sound barrier" and proceeding into the supersonic speed regime.

NACA PRELIMINARY TECHNOLOGY

As early as 1939, Ezra Kotcher, a senior instructor at the Army Air Corps Engineering School at Wright Field, advocated comprehensive flight research programs for correlating wind tunnel data with full-scale performance.[32] In 1944, over 4 years after Kotcher's recommendation, and after a name change from Army Air Corps to Army Air Force in 1941, Air Force Headquarters authorized a study into the possible development of an experimental article for investigating aerodynamic phenomena in the 600- to 650-mph range.[33] Kotcher went on to make a comparative investigation of the merits of rocket and turbojet propulsion for the transonic research aircraft in early 1944. The study showed that, from every standpoint, rocket power appeared to be better for transonic research aircraft because high speeds could be attained, making dives to high Mach numbers unnecessary.[34]

At NACA, John Stack and Eastman Jacobs were the chief proponents of a transonic research aircraft. Stack had proposed the idea in 1941 and had set up a team consisting of himself, Milton Davidson, Harold Turner, and Walter Williams to study possible research configurations. Jacobs had earlier set up a group that also included Davidson and Turner, along with Macon C. Ellis, Jr., and Clinton E. Brown. The Jacobs group was particularly interested in an aircraft with a "Campini" engine that promised ample thrust for flying at transonic speeds if aerodynamic problems could be solved.[35]

Capt. Walter S. Diehl, USN, the Navy Bureau of Aeronautics representative on the National Advisory Committee for Aeronautics, felt as early as 1942 that the research airplane appeared to be the only way to convince people that the "sonic barrier" was just a steep hill.[36] On September 22, 1944, 1st Lt. Abraham Hyatt, a Marine Corps engineer attached to the Bureau of Aeronautics, proposed the development of a turbojet-propelled high-speed research airplane for acquiring knowledge on transonic drag, flight loads, and stability and control, as well as data on engine thrust and duct inlet design.[37]

With the transonic problems of high-speed military airplanes becoming more serious by the day, representatives of the Army Air Force,

U.S. Navy, and NACA met at the NACA Langley Laboratory on March 15, 1944. Two meetings that day tied together the separate interests in the possible development of a transonic research aircraft.[38] The parties present had different ideas as to what the research aircraft should be and what it should try to accomplish, but there was general agreement that some type of transonic aircraft was urgently needed.

The NACA representatives suggested a joint effort for constructing a turbojet-powered research airplane,[39] and NACA followed up on July 10, 1944, by submitting a turbojet research airplane design proposal to the Army Air Force. The Air Force rejected the NACA design as being too conservative because its studies had shown that the turbojet engine would not give the aircraft good high-speed capabilities.[40] The NACA design was more in line with the research aircraft requirements that were recommended by the Navy's Lt. Abraham Hyatt a few months later.

As the end of 1944 approached, both the Army Air Force and the Navy had programs under way for developing their own separate transonic research aircraft. Both services relied on the advice of NACA in establishing and guiding their programs. The Navy chose to develop a turbojet research airplane, which became the Douglas D-558. The Army Air Force chose to develop the more radical rocket-propelled design—the Bell XS-1—rejecting NACA's recommendation of air-breathing engines.[41] Note that neither the XS-1 nor the D-558 resulted from extensive design competitions for finding the best transonic research airplane. Ezra Kotcher had some difficulty interesting Army Air Force contractors in developing this "one of a kind" airplane, and was rejected several times before Bell agreed to develop the XS-1. Similarly, the Navy went to Douglas, a principal Navy airplane builder at the time, and requested that the company work on the design that was to become the D-558. Careful attention by the Army Air Force, the Navy, and NACA kept the two separate research airplane programs from being duplicative. Although different in organizational structure and function, the three groups worked together in a common mission to solve the problems associated with transonic flight.

BREAKING THE SOUND BARRIER

As dawn approached on October 14, 1947, the Muroc Dry Lake in Southern California became a beehive of activity. The focus of attention was the Bell Aircraft Corporation's XS-1, America's first transonic research aircraft. This was the day that the XS-1, piloted by Capt. Charles E. Yeager of the newly named U.S. Air Force, would attempt to fly through the mysterious "sound barrier." The assembled engineers, technicians, and pilots anticipated success because the aircraft had nearly reached supersonic speeds on its previous flight 4 days earlier.

Poised and ready for its most important mission, the XS-1 was the result of almost 3 years of intensive effort by the Bell Aircraft Corporation, Reaction Motors, Inc., the Air Force, and NACA. Only 1 year

First supersonic airplane—the U.S. Air Force Bell XS-1 experimental airplane.

27

after Ezra Kotcher secured a commitment from Bell to design and construct three transonic research airplanes, Bell completed the first XS-1. Less than 16 months after construction, the XS-1 had successfully completed its first powered flight with the new Reaction Motors rocket engine. Six months later, the XS-1 was ready to go supersonic.

NACA engineers and aerodynamicists participated with the Bell Aircraft Corporation and the Air Force in many elements of the XS-1 design and flight-test program. Floyd Thompson, Assistant Chief of Research at the NACA Langley Laboratory, decided to use a "thin" wing based on transonic wingflow test results obtained by NACA's Robert Gilruth.[42] John Stack recommended that the horizontal tail be made relatively thinner than the wing to ensure that the airplane could be controlled even if critical conditions occurred over the wing.[43] Stack, Gilruth, and others further suggested that the horizontal tail be "all-moving" to ensure adequate control during transonic flight and that the horizontal tail be placed as high as possible on the vertical fin to position it above the wing wake.[44] NACA engineers set the ultimate design load factor[45] and recommended the research instrumentation that would go on the airplane.[46] Walter Williams, a NACA Langley engineer who had been on John Stack's transonic design team, was project engineer for the NACA group that was responsible for the flight tests, data gathering, and data dissemination.

Although the XS-1 had flown successfully during the contractor program and the early Air Force flights, some people believed that the airplane would disintegrate because of compressibility effects at sonic speeds. However, after his epoch-making flight of October 14, 1947, here is what Capt. Yeager had to say about it:

> With the stabilizer set at 2° the speed was allowed to increase to approximately .98 to .99 Mach number where elevator and rudder effectiveness were regained and the airplane seemed to smooth out to normal flying characteristics. This development lent added confidence and the airplane was allowed to continue to accelerate until an indication of 1.02 on the cockpit Mach meter was obtained. At this indication the meter momentarily stopped

and then jumped to 1.06 and this hesitation was assumed to be caused by the effect of shock waves on the static source.[47]

Yeager had attained an airspeed of 700 mph, Mach 1.06, at an altitude of approximately 43 000 feet. He had become the first pilot to successfully exceed the speed of sound. There had been no violent buffeting or wrenching of the airplane, and Yeager did not note anything spectacular about broaching the "sound barrier." [48]

Yeager's flight laid to rest the myth of the "sound barrier" and opened the door to the possibilities of supersonic flight. Data obtained by NASA pilots on the XS-1 and the Navy's D-558 transonic research aircraft provided a rich trove of transonic and supersonic flight information. During the period September 25, 1947, to November 6, 1958, NASA pilots made 358 research flights in various versions of these two aircraft.

Navy transonic research airplane — the Douglas Aircraft Company D-558.

SUPERSONIC CRUISE TECHNOLOGY

Another important aspect of the XS–1 and D–558 programs was the impetus they provided to the development of better ground-based research facilities and testing methods. The requirements for the best possible data in the development of these two aircraft led directly to several important innovations in wind tunnel testing. In 1945, NACA Langley engineers, adopting Gilruth's wing-flow techniques, developed the "transonic bump" method for wind tunnel testing. With the use of a carefully designed "bump," a region of transonic flow could be generated in the tunnel even though the main flow remained subsonic. Langley engineers also developed an entirely new method of testing high-speed models using special sting supports with internal balances. Using small models and the nonchoking sting support system, Langley wind tunnel engineers could test the XS–1 and D–558 almost up to the speed of sound.[49]

Perhaps the most important ground-test development spawned by the XS–1 and D–558 programs was the "slotted wall" transonic wind tunnel. In 1946, Ray Wright at NACA Langley suggested the use of slots in wind tunnel walls to more nearly duplicate free-air conditions. Making use of this concept, John Stack and his associates developed a "slotted wall" transonic wind tunnel that was essentially free of the "choking" experienced in existing solid-wall tunnels. With this tunnel it was possible, for the first time, to increase tunnel airspeed continuously through Mach 1 by merely increasing fan speed. For this important development in aviation research, John Stack and his associates at Langley received the prestigious Collier Trophy in 1951.[50]

The American quest to break the "sound barrier" also led to several revolutionary aerodynamic discoveries that have had a lasting effect on the design of high-speed aircraft. Two of the most important of these discoveries were the swept wing and the "area rule," both of which were made at the NACA Langley Laboratory.

Before 1945, military and civilian airplanes made use of straight wings that were essentially perpendicular to the direction of flight—an adequate wing orientation for speeds up to 300 to 400 mph. As airplane velocities approached the speed of sound, however, the straight wing gave rise to large increases in the drag or resistance of the airplane. In 1944, Robert T. Jones of NACA Langley noted that these drag increases could be substantially reduced by sweeping the wings back from the direction of flight. Jones had developed his wing-sweep theory independently and without knowledge of the pioneering wing-sweep effort of German scientist Adolph Busemann in 1935. The discoveries of these two scientists changed the shape of high-speed airplanes, and essentially all such aircraft now employ swept-back wings.

The discovery of the "area rule" was almost directly due to the development of the "slotted wall" transonic wind tunnel and the capability it provided to test models near Mach 1. In 1951, while conducting such tests, NACA Langley engineer Richard Whitcomb and his team observed the expected shock disturbances from the nose of a model, but also found additional disturbances emanating from the trailing edges of the wings. Whitcomb believed that these latter disturbances were leading to high drag and that they were caused by irregularities in the cross-sectional area distribution at the wing/fuselage juncture. He theorized that drag could be substantially reduced if the total combined cross-sectional area of the wing, fuselage, and tail were adjusted to approach that of an ideal streamlined body. Thus, the fuselage should be constricted where the wings were attached and expanded at their trailing edges. Subsequent wind tunnel tests proved Whitcomb's concepts, and the transonic "area rule," with its attendant "coke bottle" fuselage shape, was born.[51] The "area rule," which resulted in a Collier Trophy for Whitcomb, was later expanded to supersonic speeds by Boeing and NASA engineers and is a major tool in the design of high-performance supersonic airplanes.

Thus, the necessary first step in the evolution of supersonic cruise flight — "breaking the sound barrier" — had been successfully taken. As NACA engineers and scientists participated in this effort, they produced breakthroughs that would have far-reaching effects on the shape of future civilian and military aircraft. However, NACA was aware that short-duration supersonic flights of small experimental airplanes did not ensure the ready development of long-range supersonic aircraft. Much work was yet to be done.

During the latter part of the 1940s and the first half of the 1950s, NACA prepared for the new era of supersonic flight. With the support of the U.S. Congress, a wind tunnel construction program was initiated at all three of NACA's ground test centers. As a result of the Unitary Plan Act, passed by Congress on October 27, 1949, a large transonic/supersonic wind tunnel complex was built at NACA Ames, Moffett Field, California; a new supersonic tunnel was constructed at NACA Langley, Hampton, Virginia; and a large supersonic tunnel, dedicated to propulsion system integration, was established at NACA Lewis, Cleveland, Ohio. In addition, the Unitary Plan Act authorized a new Air Force aeronautical test center at Tullahoma, Tennessee, now known as the Arnold Engineering Development Center (AEDC). Under the Unitary Plan, NACA was to concentrate on industrial development work (i.e., commercial aircraft), and the Air Force AEDC was to focus on military development testing.[52]

In addition to acquiring these facilities with increased research capability, NACA made other preparations for the approaching era of supersonic cruise flight. At the Ames and Langley aeronautical centers, scientists began to collect data on a myriad of aeronautical configurations that appeared to be suitable for supersonic cruise flight. These two NACA groups also began the development and validation of analytical methods that could be used in supersonic design and analysis. The Lewis propulsion center initiated a massive effort for developing turbojet engine technology for supersonic airplanes. The effort covered the aerodynamic design of all components of the engine, inlet, and nozzle and the development of new high-temperature materials and cooling techniques for turbine blades and considered the most suitable design characteristics of a turbojet engine. Meanwhile, at the NACA Flight Research Center at Edwards, California, engineers and pilots continued to collect valuable transonic and supersonic data on a bevy of high-speed research airplanes such as the X-2, X-4, X-5, D-558-II, and XF-92. Also under way at NACA was an advanced structures research program in which new materials, such as titanium, and the "super alloys" were being investigated at the temperature levels to be expected during supersonic flight.

The preparations for the development of long-range supersonic airplanes were being made. The question now was: When would the development begin?

CHAPTER 4

NACA/NASA Supportive Supersonic Cruise Technology (1956–1971)

As NACA prepared for the approaching era of supersonic cruise flight, the military services followed up quickly on the knowledge gained in the X-1 and D-558 programs. By 1953, the first American supersonic fighter, the North American F-100, had been developed, and it was soon followed by the Mach 2 fighter, the Lockheed F-104. In 1956, the Convair B-58, a medium range bomber also having Mach 2 "dash" capability, was brought into the Air Force inventory.

All of these aircraft performed well in the transonic and supersonic speed regimes, but they all had rather short supersonic range and duration of flight. Even though a number of aircraft were now capable of flying at supersonic speeds, the total flight experience at supersonic speeds was still very small. For example, during its first 10 000 hours of flight time, the B-58 bomber logged only 500 hours at speeds above Mach 1.5.[53]

This early flight experience with supersonic aircraft gave little indication that a vehicle could be developed that would be capable of sustained long-range supersonic flight. At this point, it appeared that the advent of the concept of supersonic cruise flight would hinge on some substantial breakthrough in aerodynamic, structural, or propulsive efficiency.

THE U.S. AIR FORCE B-70 BOMBER PROGRAM

In late 1951, Gen. Curtis LeMay, Commander of the U.S. Air Force's Strategic Air Command, began looking around for a B-47 bomber replacement. A "split mission" aircraft was envisioned — one that would cruise subsonically to the enemy early-warning line (EWL) and then "dash" supersonically to the target. On October 14, 1954, after several years of study and consideration, the Air Staff published a general operating requirement (GOR) for a supersonic strategic bomber for the 1965-1975 time period.[54] This GOR initiated the preliminary studies that would lead to the Air Force supersonic B-70 bomber program.

Four companies responded to the Air Staff GOR and undertook preliminary design studies for such an aircraft: General Electric for propulsion, IBM for homing and navigation devices, and Boeing and North American for design integration and the manufacture of the entire weapon system.[55]

At the time of these preliminary B-70 studies, the Air Force perhaps expected little more than a larger version of the B-58. There was little to indicate that more than supersonic "dash" capability could be achieved, and "the only recourse available to the design engineer was the so-called brute-force concept — tremendous fuel capacity and an optimized aerodynamic shape for the target sprint."[56]

General LeMay is said to have groaned when he saw the resultant 1955 B-70 design; the concept called for jettisonable wing-tip tanks that weighed 191 000 pounds and were about the size of a B-47. Grossing out at more than 750 000 pounds, the design featured a huge canard arrowhead control surface that cut the pilot's forward vision by 50 percent.[57]

Shortly after Air Force rejection of this preliminary concept, a NACA aerodynamic "breakthrough" was revealed that had dramatic implications on the B-70 program and on future supersonic design philosophy. This breakthrough was the product of A. J. Eggers and C. A. Syvertson of the NACA Ames Research Center.[58] Their research indicated that a substantial quantity of favorable "compression lift" could be created on the under surface of a wing at supersonic speeds by the proper placement of a body or "splitter" beneath the wing surface. A supersonic airplane that employed this concept could effectively ride its own shock wave. The "compression lift" concept was tested at the NACA Ames, Lewis, and Langley centers. Dr. Hugh Dryden, the NACA Director, had this to say about the compression lift concept in 1958 testimony to the U.S. Congress:

> A year or so ago, the Air Force was reconciled to the idea that the best it could obtain in the way of performance from a new large bomber would be what the engineers call high subsonic cruise plus supersonic dash similar to the B-58. About a year ago a strange and wonderful thing happened. It was as if the pieces of a jigsaw puzzle began falling into place. Almost simultaneously research programs that had been under way at NACA labs in Virginia [Langley], California [Ames], and Ohio [Lewis] began paying off.

> The result — this is oversimplification, but it is not overstatement — was that the companies [i.e., North American and Boeing] and the Air Force suddenly realized it would not be much harder to design a long-range bomber that could fly its whole mission supersonic than to design one that would fly subsonic all the way with only a small fraction of the flight supersonic. Not only that, but the top speed of the prospective bomber was raised to Mach 3, about 2000 miles an hour.[59]

The aerodynamic promise of Eggers and Syvertson's compression lift concept encouraged the Air Force to proceed with the B-70 bomber program. A North American configuration, which made use of this concept, was selected as the winning entry in the B-70 competition. The company was awarded a design and production contract on December 23, 1957. The production contract was not to remain in effect very long, however. On December 3, 1959, a teletype from the Pentagon informed North American that "THE B-70 IS BEING REDIRECTED AS FOLLOWS: (A) CEASE ALL STUDY, DESIGN, DEVELOPMENT AND FABRICATION AND TEST WORK TOWARDS THE B-70 WEAPON SYSTEM IN WING STRENGTH IN ACCORDANCE WITH THE MASTER PHASING SCHEDULE; (B) PROCEED WITH DESIGN, DEVELOPMENT, FABRICATION AND TEST WORK AND PRODUCE ONE XB-70 FOR THE EARLIEST POSSIBLE FLIGHT DATE CONSISTENT WITH TENTATIVE FUNDING CEILINGS. . . ."[60]

North American B-70 bomber program spurred supersonic cruise research.

This Pentagon directive was eventually amended to permit the construction of two B-70 aircraft, but it effectively signaled the end of the U.S. military's interest in long-range supersonic cruise aircraft. The intercontinental ballistic missile (ICBM) had replaced the manned bomber as America's major strategic weapon system.

The first of the two B-70 prototypes authorized by the Pentagon made its initial flight on April 21, 1964. It attained the design cruise speed of Mach 3 (2000 mph) in May 1965. The second aircraft made its first flight on July 17, 1965. This latter aircraft had been heavily instrumented by NASA for research experiments when it was lost in a tragic midair collision with an F104 on June 8, 1966. The two B-70 prototypes flew a combined total of about 100 missions.

The B-70 bomber was a remarkable airplane for its time period. Designed before the development of high-speed computers and automated aerodynamic and structural design methods, the B-70 provided a mass of useful data on aerodynamics and sonic boom. NACA provided one of the principal design features of the airplane (i.e., compression lift). Although it was never conclusively proven whether the

B-70 obtained the full predicted benefits of compression lift during flight, the principle of compression lift or "favorable lift interference" was incorporated in later aerodynamic design methods.

THE U.S. SUPERSONIC TRANSPORT PROGRAM

There was a feeling in some quarters that the cancellation of the Air Force B-70 bomber program would be a mortal blow to the concept of supersonic cruise flight, at least in America. The B-70 was expected not only to establish the feasibility of such flight, but also to serve as a proving ground for a future commercial supersonic transport. A wing of long-range B-70 bombers could rapidly obtain the 20 000 or so hours of supersonic flight experience that were desired prior to commercial development. Moreover, later models of the B-70 could try out new technology for later application on the commercial transport airplane.

Many were happy to see the demise of the B-70 production program and were even more pleased that this demise might inhibit or deter the development of a commercial SST. Concerns were rising about the effect of a fleet of SSTs on the environment, and the questions about pollution, safety, radiation, noise, etc., were beginning to dim some of the ardor for the apparently insatiable quest for higher speed. One of the most disturbing phenomenon associated with supersonic flight was the sonic boom—a startling thunderclap that is produced by a vehicle moving at supersonic velocity through the atmosphere.

NASA, by 1960 heavily involved in the national space program, looked on the challenge of supersonic cruise flight as well within its chartered responsibility for preserving the leadership of the United States in aeronautical science and technology. While recognizing the difficult problems that had to be solved before commercial supersonic flight, NASA saw such flight as a rational and necessary evolution in commercial transportation. Accordingly, NASA began to build rapidly on its base of supersonic technology with the firm belief that this technology would be useful to the United States. This belief was strongly shared by others both within and without the government.

In the early 1950s, NACA had initiated concerted efforts to improve the technology of supersonic cruise flight. Now, in the latter half of the decade, the newly formed NASA was stepping up the research and technology efforts. Fundamental programs were under way in all design and operational areas associated with supersonic cruise flight. At each of the NASA aeronautical centers, engineers were seeking ways to provide the levels of aerodynamic, propulsive, and structural efficiencies that would be required of a viable commercial SST. A variety of aerodynamic and propulsive concepts that had the potential for meeting the required efficiency levels were exhaustively tested in the NASA low-speed and new supersonic Unitary Plan wind tunnels. Programs for studying the air traffic control and other operational problems were initiated in cooperation with the FAA. In 1960, NASA formed a Supersonic Transport Research Committee to coordinate and guide this mushrooming supersonic technology effort.

In mid-1960, the NASA Langley Research Center summarized the effects of supersonic research of the 1950s on the technical status of the supersonic transport. John Stack, who had been heavily involved in NACA's earlier transonic research, said in his introduction to this technical summary:

> The prospects for commercial flight at supersonic speeds herald a new era in the transportation field. The successful development of a supersonic transport is of vital importance to the national prestige as well as the commercial stature of the United States. If the United States is to achieve a supersonic air transport capability at the earliest practical date, a vigorous effort is demanded on all fronts.[61]

Stack went on to comment about the potential SST configurations that had evolved from NASA studies: "While many of these configurations have serious limitations for commercial operation, reasonably clear definitions of the problem areas have been achieved and possible new approaches are under study." [62]

NACA/NASA SUPPORTIVE TECHNOLOGY

Therefore, in 1960, NASA believed that the development of a commercial SST was of vital importance to the United States and that supersonic research had come a long way toward providing the technology for a viable SST, but that much research effort was still required before the goal could be reached. NASA was busily conducting this research.

In 1961, the preliminary groundwork for establishing a U.S. SST program moved forward. Representatives of NASA, the Federal Aviation Agency, and the Department of Defense met to consider the responsibilities of each agency in the event that such a program would be approved by the U.S. Congress. Out of this meeting came a joint report which indicated that "A vigorous [SST] effort must be started immediately in order to have an operational aircraft in the 1970 time period." [63] The report suggested the following agency roles in such an effort:

- Federal Aviation Agency — Leadership and fiscal support
- Department of Defense — Administrative and technical support by the U.S. Air Force
- National Aeronautics and Space Administration — Basic research and technical support.

Secretary of Defense Robert S. McNamara, NASA Administrator James E. Webb, and FAA Administrator N. E. Halaby signed this joint agency report. The three agencies established an SST Steering Group, consisting of the Administrator, FAA, Chairman (Halaby); the Assistant Secretary of the Air Force for Research and Development (Joseph S. Imirie); and the Director of Aeronautical Research, NASA (Stack). The Steering Group was to coordinate joint efforts on SST problems.

Shortly after this joint agency report was signed, an FAA request for $12 million in fiscal year 1962 SST funding came before the U.S. Senate. This funding was to provide for FAA-managed research that would augment the supersonic effort at NASA. On July 31, 1961, a motion was made in the Senate to strike the SST funding from the FAA budget. The first SST funding measure barely survived when the "motion to strike" the funding failed to pass on a 35 to 35 tie vote.[64] An SST funding level of $11 million was subsequently approved by Congress, and the FAA research program was under way.

FAA received an additional $20 million in fiscal year 1963 to continue the preliminary SST research program. NASA assisted FAA in monitoring the research contracts and grants that evolved from this program.

Meanwhile, NASA engineers were homing in on some aerodynamic configurations that showed very good potential for meeting the requirements of a commercial SST. From approximately 20 concepts, four aircraft arrangements had emerged as likely SST candidates after a series of analyses, wind tunnel tests, and refinement. In the latter half of 1962, NASA prepared to submit these four concepts to an SST feasibility study by the American airplane industry.

NASA SUPERSONIC COMMERCIAL AIR TRANSPORT (SCAT)
FEASIBILITY STUDIES

The NASA/Industry SCAT Feasibility Studies were initiated in January 1963, making use of $1 million of NASA funds. The major purposes of the studies were to bring the NASA SST effort into focus and to establish the technological state of the art for guidance of a possible national SST program. NASA would also receive the unique expertise of industry in the evaluation of the four candidate SST concepts and in the identification of areas requiring further research attention.

The four NASA configurations submitted for industry feasibility studies bore the designations SCAT-4, SCAT-15, and SCAT-16 (generated at the Langley Research Center), and SCAT-17, which was developed at the Ames Research Center. All three of the Langley designs incorporated highly swept arrow wings. The SCAT-4 wing was fixed, the SCAT-15 made use of auxiliary variable-sweep panels to improve low-speed performance, and the SCAT-16 was a classical variable-sweep wing design. The Ames SCAT-17 configuration was characterized as a fixed-delta-wing airplane with a small auxiliary wing or canard near the nose.

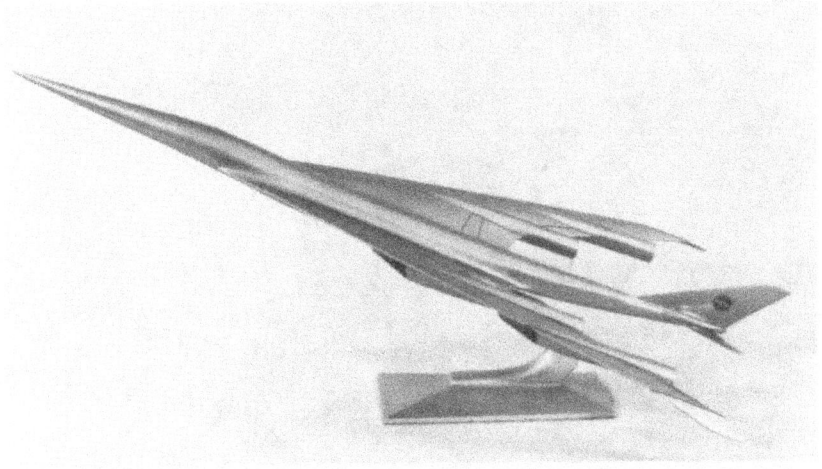

NASA supersonic transport feasibility concept — the highly swept fixed arrow-wing concept SCAT-4.

NASA SST feasibility concept — the variable-sweep SCAT-15 with blended body and interlocking wings (forerunner of the fixed-wing SCAT-15F.)

NASA SST feasibility concept — the SCAT-16 with highly swept arrow wing in aft position.

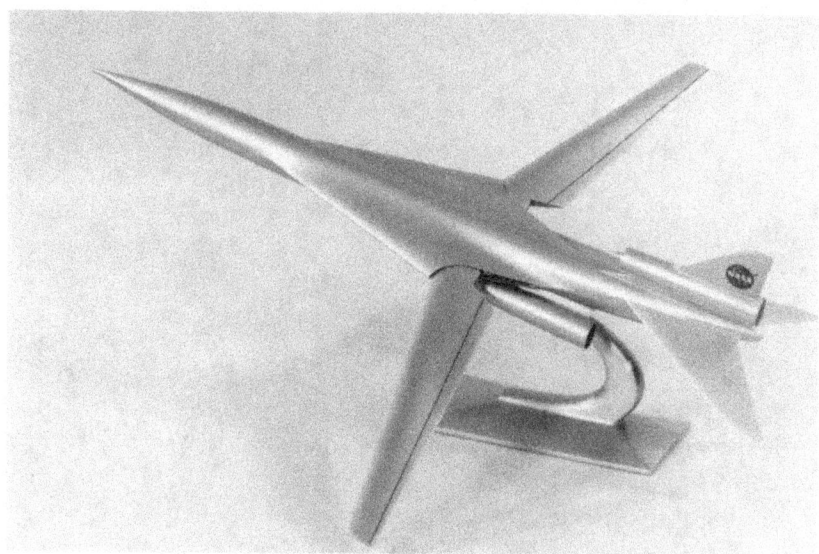

NASA SCAT-16 with wings in swept-forward position.

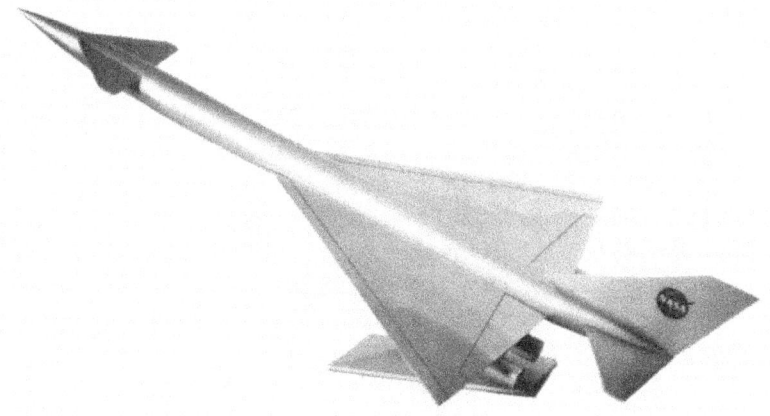

NASA SST feasibility concept — the fixed delta-wing SCAT-17 with forward canard and aft tail.

The Boeing Airplane Company and Lockheed California Company were selected to assess the feasibility of the four SCAT designs on a 3200-nautical-mile SST mission with 125 passengers at a cruise Mach number of 3, as well as the feasibility of one of the designs as an aluminum Mach 2 transport. In addition, they were to evaluate several engine types and levels of propulsion technology. In flying the mission, the sonic boom levels could be no more than 2 pounds per square foot (psf) in climb and no more than 1.5 psf while at cruise conditions. The engine noise would be comparable to present-day subsonic jet aircraft.

The results of the NASA SCAT Feasibility Studies were reported on September 17–19, 1963, at a Langley Research Center conference.[65] The studies indicated the following:

- Either a Mach 3 variable-sweep transport, patterned after the SCAT-16, or a fixed-wing transport, following the lines of the SCAT-17, could make the New York/Paris mission with 125 passengers.
- If the SCAT is to be a competitive transport, an advance in the propulsion state of the art would be required.

- A Mach 2 aluminum SCAT could not make the mission within the sonic boom constraints used in the study.
- Titanium was preferred over stainless steel as the basic material for a Mach 3 SCAT.
- For the transport to have satisfactory characteristics, research in several areas would be urgently needed. The aircraft needs to have better aerodynamic performance than that used in the studies, and configurations need to be devised that can obtain the high aerodynamic performance while still maintaining low structural weight. Research and development work in the engine field is required. Research is needed in stability and control to improve flying qualities of the aircraft. Considerable work is needed to establish confidence in the use of titanium. Research is needed to determine if the sonic boom can be reduced. Research is needed to determine the public acceptance of sonic boom so that boom restrictions can be set at the highest possible level.

In essence, on the basis of then-current status of technology in almost all areas of research, the NASA SCAT Feasibility Studies of 1963 indicated that a commercial supersonic transport was not feasible. However, 1 month later, on August 15, 1963, FAA circulated to industry the "Request for Proposals for the Development of a Commercial Supersonic Transport." The United States SST program was under way.

FAA SST Evaluations

While the NASA SCAT Feasibility Studies were in progress, President Kennedy and the U.S. Congress took the actions that launched the American commercial SST program. With the release of the FAA's request for proposals, NASA's role changed from one of having its own concepts evaluated by the airplane industry to one of evaluating the SST concepts of the airplane industry.

The FAA allowed the interested airplane and propulsion companies about 4 months to develop and submit proposals for the SST development program. During this time, NASA participated with the FAA in setting up evaluation criteria and organizing the SST Evaluation Team. The FAA development plan called for one evaluation during January 1964, which was to be followed by:

- Selection of the winning airframe/propulsion combination by May 1964
- A 1-year detailed design effort ending in May 1965
- First flight of a United States SST in the fall of 1968
- SST entry into passenger service in the summer of 1970 [66]

As it turned out, four FAA SST evaluations were held before the selection of the "winning combination" on December 31, 1966. The evaluation process extended from January 1964 to October 1966 — a period of 33 months. During this time, NASA provided totals ranging from 41 Evaluation Team members in the first evaluation to 74 members in the last evaluation in September 1966. Approximately one-third of the evaluators were from NASA, and about two-thirds of NASA's representation were from the Langley Research Center. Langley engineers also provided a large part of the analytical tools that were used in the evaluation process and, during the last three evaluations, conducted extensive wind tunnel tests on models provided by the contractors. In the final evaluation, Ames Research Center conducted low-speed wind tunnel tests of contractor models.

Three airframe companies — Boeing, Lockheed, and North American — and three engine companies — General Electric, Pratt and Whitney, and Curtis Wright — submitted proposals for evaluation in January 1964. The first FAA/SST evaluation found that none of the airframe/engine combinations met the basic criteria of range (4000 statute miles), passengers (125 to 160), or sonic boom (2.0 psf during climb, 1.5 psf during cruise). The principal problem areas appeared to be cruise aerodynamic performance and sonic boom.

From this first evaluation, Boeing, Lockheed, General Electric, and Pratt and Whitney proposals were judged to have the most potential, and they were continued in a competition that was to last nearly 3 years. In none of the subsequent FAA evaluations, including the final one in late 1966, did any of the airframe/engine combinations simultaneously meet all of the evaluation criteria. Throughout the selection process, Boeing steadfastly stuck to the variable-sweep concept with its uncertain weight penalty and promised advantages in mission versatility and cruise performance. Lockheed just as firmly stayed with their relatively uncomplicated fixed double-delta-wing design, which offered simplicity at the expense of possible growth potential. Because each concept had possible advantages and potential disadvantages, the selection of the Boeing/General Electric (B-2707-200/GE-4) combination on December 31, 1966, was less than a clear-cut decision.

The difficulty of this decision was borne out in April 1968 when Boeing dropped the variable-sweep B-2707-200 and changed to the B-2707-300, a fixed double-delta-wing design not unlike the Lockheed configuration. After Boeing had worked the Dash-300 for most of 1968, the FAA called for a validation team to review the design. The United States Supersonic Transport Integrated Configuration Validation Group met on-site at Boeing in December 1968 and January 1969 to review the Dash-300 design. This validation, unlike the previous rigorous FAA evaluations, represented little more than an audit of Boeing substantiating information.

This "validation" exercise was the last formal contact between the NASA evaluators and the FAA SST program before the cancellation of the program in May 1971. Individual contacts of NASA engineers with the Boeing SST team continued until the program was terminated.

The cancellation of the U.S. SST program was a severe blow to those who had devoted a large part of their careers to research on supersonic cruise flight. The aura of negativism and controversy that surrounded the program in its last phases tended to depreciate the real accomplishments that had been made by both NASA and the SST contractors. Some of these accomplishments are summarized herein.

IMPROVEMENTS IN SUPERSONIC DESIGN AND ANALYSIS METHODS

At the beginning of U.S. interest in supersonic cruise flight, the aeronautical literature contained theories for optimum supersonic area distributions, optimum supersonic wings, and optimum supersonic lift

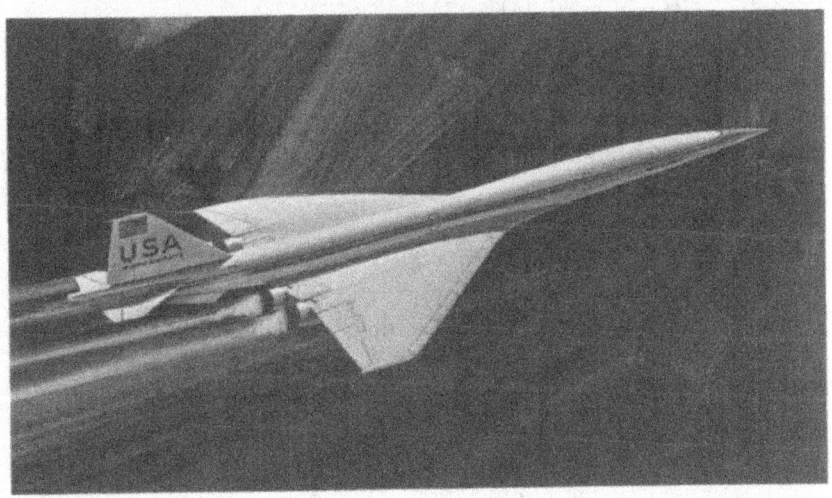

Boeing B2707-300 with four GE-4 engines — the proposed U.S. SST at the time program was canceled.

distributions.[67, 68] The problem was that no means were available for designing these optimum aerodynamic features into a complicated supersonic aircraft. There were also no "real flow" constraints on the first-order theories and no means for handling the design effects of "favorable lift" interference.

Immediately preceding and during the course of the U.S. SST program, NASA and industry engineers adapted the principal elements of these basic theories into high-speed digital computer programs. "Real flow" constraints and "favorable lift" interference were incorporated, and each element of the theory was validated with carefully controlled wind tunnel tests. These programs can be used to design, analyze, and optimize the aerodynamic characteristics of complex, arbitrary supersonic configurations with a high degree of accuracy.

The design and analysis methods developed during this program were the standards for supersonic evaluation of the proposal SST configurations, and they are widely used today in the design of supersonic military airplanes.

49

SUPERSONIC CRUISE TECHNOLOGY

IMPROVEMENTS IN SONIC BOOM TECHNOLOGY

Prior to NASA's emphasis on supersonic cruise technology, little was known about sonic booms except that they were generated by airplanes flying at supersonic speeds. The principal airplane factors that caused them and the manner in which they were propagated were still mysteries.

Using the asymptotic pressure-field theory developed for supersonic projectiles[69] and the lift-area equivalency principles of the supersonic area rule,[70] NASA Langley developed and validated a "far field" sonic boom estimation procedure for use in the FAA SST evaluations. It was later discovered at Langley that the SST, because of its relatively large size, would generate "near field" sonic booms with somewhat lower levels than those predicted by the "far field" method. The discovery of this "near field" application to the SST opened up a new field of research in sonic boom minimization. This application was immediately incorporated into the methodology for sonic boom prediction and has been the subject of many research papers.

Until this point, the sonic boom disturbance was assumed to be propagated through a mean uniform atmosphere that accounted for differences between the atmospheric pressures between the airplane and the ground. In 1969, this approximation was greatly improved with the development of a method for propagating the sonic boom through a stratified atmosphere.[71]

In addition to these developments in sonic boom methodology, NASA engineers participated in the measurement of sonic booms from essentially every supersonic airplane, participated in all of the FAA sonic boom acceptability programs, gathered a bulk of data on the effect of atmospheric disturbances, and conducted studies on the possible damages that might be caused by sonic booms. Thus, in the period 1959–1969, NASA brought the status of knowledge on sonic booms to a very high level.

IMPROVEMENTS IN SUPERSONIC CRUISE AERODYNAMIC EFFICIENCY

The lift/drag ratio is used by aerodynamicists as a measure of the aerodynamic efficiency of a given configuration. At the beginning of

50

the U.S. SST program, lift/drag ratios of 7.0 to 7.5 were typical of aircraft designed for supersonic cruise flight in the range from Mach 2 to Mach 3.

As one means of validating the supersonic design methodology that had recently been developed, in 1964 Langley engineers designed an SST configuration that demonstrated a lift/drag ratio of 9.3 at Mach 2.6. This level was 25 to 30 percent above the previous state of the art. The design, called the SCAT-15F, was a fixed-wing version of the variable-sweep SCAT-15 that was considered in the earlier SCAT Feasibility Studies. The SCAT-15F incorporated the principles of wing design, wing/fuselage integration, engine placement, and favorable lift interference that were validated in the process of developing the supersonic design methods.

The SCAT-15F was considered by both Lockheed and Boeing during 1965 and was studied again by Boeing in 1968 as a backup to the Dash-300. The concept was also the subject of a 1966 NASA study on the feasibility of a low sonic boom domestic transport requested by the FAA.

Although neither contractor adopted the SCAT-15F as a primary SST concept in spite of its superior supersonic cruise efficiency, many of the design features of the aircraft were incorporated in the contractor designs.

IMPROVEMENTS IN PROPULSION TECHNOLOGY

By the end of the U.S. SST program, turbojet and turbofan engines producing 50 000 to 60 000 pounds of thrust appeared feasible compared to about 30 000 pounds at the start of the program. On October 28, 1966, the GE-4 engine produced 52 600 pounds of thrust on the test stand, and on December 11, 1966, the other SST engine, the Pratt and Whitney JTF17, produced 57 000 pounds of thrust — a Free World record at that time.[72]

Much of the improvement in supersonic engine capability was attributable to the high-temperature turbine and advanced materials

51

NASA SCAT-15F fixed-wing concept that advanced state-of-the-art levels of supersonic cruise aerodynamics.

programs of the NASA Lewis Research Center. This same technology has also brought improvements in present-day subsonic turbojet and turbofan engines.

Kelly Johnson, the noted Lockheed designer of the U-2, SR-71, and other advanced airplanes, once said, "At supersonic flight speeds around Mach 3, our current and proposed jet engines produce only about 25 percent of the thrust propelling the aircraft. The inlet duct pressure distribution provides about 50 to 60 percent of the thrust and the exhaust ejector the rest." [73]

The importance of the propulsion system inlet was not overlooked in NASA's effort in support of the SST program. The NASA Ames Research Center became the hub of the SST inlet program and provided much of the technology that would go into the Boeing transport. Engineers at Ames conducted detailed large-scale tests of axisymmetric and two-dimensional inlets in the Ames facilities and developed a number of innovations to improve inlet performance.

IMPROVEMENTS IN STRUCTURES AND MATERIALS TECHNOLOGY

The U.S. decision to develop a Mach 2.7 transport led to the requirement for a new structural material. Aluminum, the material most generally used in airplane construction, would not survive in the 500°F temperature environment experienced at that speed. Although stainless steel had been used in the construction of the B-70 supersonic bomber, the scant design information available indicated that titanium would be a better high-temperature material. A large part of NASA's effort in support of the SST was devoted to expanding the status of knowledge on titanium. At both the Langley and Lewis Research Centers, exhaustive research programs were conducted to determine the design variables, material properties, and fabrication problems of titanium. Now, because of these technology efforts and other studies carried out by the aerospace industry, titanium has replaced aluminum in some areas of current subsonic airplanes, even where temperature is not a problem.

As in the aerodynamics area, the design methodology in structures was vastly improved because of the SST program. The more complex structural design problems associated with the SST — compared to subsonic aircraft — led to the requirement for advanced methods for design and analysis. As a result of efforts by both NASA and industry, sophisticated and accurate computerized structural design and analysis methods were developed. Adaptations of these SST methods are currently used in the design of advanced subsonic aircraft and are being applied to automotive and other vehicle designs.[74]

OTHER TECHNOLOGY ADVANCES DERIVED FROM THE
U.S. SST PROGRAM

The concepts of relaxed static stability and variable camber flaps on the wing leading edge were developed and evaluated in the U.S. SST program and have since been applied to the F-16 fighter. Various elements of the Boeing 747 jumbojet are direct descendants of development work on the SST. Digital displays and advanced navigation systems developed for the SST are now being used in the advanced subsonic jet transport—the Boeing 767.[75]

Stratospheric emission impact is another important technology area that was spurred by the SST program. Although particulate emission is not a problem unique to the SST, unknowns about the problem contributed to program cancellation. The SST program, in turn, perhaps led to the U.S. Congressional mandate to sort out the unknowns about atmospheric pollution. From this mandate grew the Department of Transportation's Climatic Impact Assessment Program (CIAP) and the High Altitude Pollution Program (HAPP). In addition to the contributions of its own atmospheric analysis and monitoring efforts, NASA was a prominent participant in the interagency activities arising from CIAP and HAPP. As indicated later, these programs led to a much better understanding of atmospheric pollution phenomena.

The cancellation of the U.S. SST program perhaps kept these important technology advances from being a complete victory for those who had worked on them. Would the same "hollow" victory be the fate of participants in the foreign SST programs?

CHAPTER 5

Foreign Supersonic Transport Programs

Supersonic transport programs by the Soviet Union and a consortium of the British and French governments perhaps precipitated America's entry into the SST race. Problems experienced in these foreign programs might also have influenced America's decision to step out of the race, at least for the present. Whatever the case, the cancellation of the U.S. effort in 1971 left the supersonic market to the Soviet TU-144 and the British/French Concorde. Since these foreign programs have probably directly or indirectly influenced the NASA supersonic technology effort, it might be well to take a brief look at them before proceeding.

THE SOVIET TU-144

The first flight of a prospective commercial SST was made by the Soviet Union's TU-144 on December 31, 1968. Design development of this airplane was initiated in 1962, and the formal development project was started in 1964. The airplane was designed to cruise at Mach 2.2 (approximately 1350 miles per hour) and to carry 140 passengers a distance of 4000 statute miles. The TU-144 employs a highly swept ogee-delta wing, and the production version of the airplane makes use of small retractable canard surfaces near the nose to control approach speed. In the original version, four Kuznetsov NK-144 turbofan engines in the 38 000-pound thrust class were clustered close together along the wing centerline in twin-ducted nacelles that stretched over the full length of the wing.[76] Andrei Tupolev, the well-known Soviet

Soviet TU-144 supersonic transport airplane.

Soviet TU-144 (courtesy of Aviation Week and Space Technology, McGraw-Hill, Inc.).

designer responsible for the TU-144, revealed in June 1966 that he had borrowed the B-70's " 'compression lift' design to let the wing ride on top of the shock wave induced by the engine intake splitter. . . ." [77] This technological advancement (discussed in Chapter 3) had been discovered by NACA's Eggers and Syvertson.

Although most elements in the development of the TU-144 have been clothed in secrecy, many problems were obvious in the early prototypes. Cruise-drag was a problem because of the poorly sculptured wing design and engine placement. The four engines were clustered in a split nacelle directly under the fuselage. This arrangement necessitated that the landing gear and carriage be placed in the outer wing. Consequently, a large proportion of the outer wing was unavailable for fuel.[78]

Based on engine technology available at the time, the turbofan appeared to be a poor choice of power plant for the TU-144. The average jet velocity of a turbofan engine is less than that of a turbojet of comparable technology. Consequently, to provide thrust equal to a turbojet, the comparable turbofan must be larger in diameter, or additional fuel must be burned in the fan duct. Eastern European sources indicated that the TU-144 used "30-40 percent of available reheat" at Mach 1.9, a situation that could cause severe fuel consumption rates and drastically reduce range.[79]

Because of the rather crude status of the first TU-144 prototype, it can be conjectured that this airplane was constructed to "scoop" the Concorde, which it did, and to provide data for incorporation into later prototypes. There were three such prototypes in the program leading up to the production aircraft.[80]

The "production" version of the TU-144 showed many refinements over the earlier prototypes. The wing was more elegantly tailored, and the engines were moved from the clustered arrangement under the fuselage to two twin nacelles outboard on the wing. The landing gear was stowed in more favorable locations within the engine nacelles, permitting a 40 000-pound increase in fuel capacity. The basic engines were not uprated, but larger burners were incorporated to increase thrust. The production version of the TU-144 first flew in mid-1972.[81]

The TU-144 was involved in a tragic accident at the Paris Air Show in 1973. While making low-speed runs, the aircraft went out of control and crashed, killing the crew and a number of spectators. Although no official report as to the cause of the crash has been given, it is believed that the breakup in flight was due to structural overload.[82]

The current status of the TU-144 program is difficult to ascertain. Rumors are that the aircraft was used in passenger service for a short time and in freight service for a while and is now grounded. Other rumors say that an advanced version of the TU-144 is under development. Aleksei I. Smenkov, Soviet first deputy minister of civil aviation, said after the tragic 1973 Paris accident, "We know there are plans for a larger airplane. Of course, our industry will not stop with this model. But so far, we have been talking about real airplanes, airplanes we know." [83] A lot has happened since Smenkov made this statement, but none of it has been particularly good for the concept of supersonic cruise flight. About the only thing that can be said with certainty is that, in building and flying three prototype and five or so production SSTs, the Soviet Union has amassed a great deal of experience in commercial supersonic cruise flight.

THE BRITISH/FRENCH CONCORDE

In spite of being cast as a transportation "heavy" by critics around the world, the British/French Concorde ranks as one of the foremost technical achievements that has ever been made. The two nations that developed this aircraft not only spoke different languages, but also used different measurement systems. Yet, out of this unusual alliance came the first and, so far, only commercial supersonic transport in regular passenger service. Like it or not, the Concorde is a remarkable airplane. It reduced the trip times between continents to one-half of those of the best of the subsonic jet transports, an accomplishment that would have been cheered in bygone years. The Concorde is perhaps the world's most tested transport airplane and, in its operations to date, has experienced no major accidents and has had no passenger fatalities.

The Concorde was probably born out of the desire of the British and French aircraft industries to break the virtual monopoly that the American airplane industry was building in the commercial subsonic jet transport market. The British had missed the chance to corner this market when early problems beset the De Havilland Comet, the first commercial jet transport. The British Overseas Airways Corporation (BOAC) had introduced the DH Comet I turbojet transport into passenger service on May 5, 1952, some 6 years before America's first turbojet transport, the Boeing 707. In its second year of operation, however, two Comets mysteriously exploded in flight, killing 55 people, and the Comets were withdrawn from service in April 1954.

The British Concorde SST at Dulles Airport in Washington, D.C. (courtesy of BOAC).

British Concorde (courtesy of BOAC).

59

French Concorde (courtesy of BOAC).

Subsequently, exhaustive tests indicated that metal fatigue had caused the Comet disasters. By the time the problem was fixed and the Comet returned to service on October 4, 1958, however, the American turbo-jet aircraft had just about cornered the market.[84]

The problems with the Comet prompted the British Ministry of Aviation to set up the Supersonic Transport Aircraft Committee (STAC) to consider the pros and cons of a British supersonic transport program. The first meeting of STAC on November 5, 1956, was really the beginning of the British SST effort. At this meeting, interest centered on a slender delta wing configuration for long-range Mach 2 flight and an M-wing concept with waisted fuselage for short-range operation at Mach 1.2.[85]

The STAC considered all the problems associated with supersonic flight from the sonic boom to ozone, radiation, and airport noise. When Sir Morien Morgan submitted his STAC report to the controller of aircraft at the Ministry of Supply on March 9, 1959, no less than 500 separate studies were attached. The principal recommendations of the report were that the British government should embark as soon as possible on a program for two airliners — a 3450-mile, 150-passenger airplane with a cruise speed of not less than 1200 mph (Mach 1.8) and a second airplane capable of carrying 100 passengers over a stage length of 1500 miles with a cruise speed of about 800 mph (Mach 1.2).[86] The report went on to say:

> Since this country's future will depend on the quality of its technological products and since its scientific manpower and resources are less than those of the U.S.A. and U.S.S.R., it is important that a reasonable proportion of such resources are deployed on products which maintain our technical reputation at a high level. A successful supersonic aircraft would not only be a commercial venture of high promise but would also be of immense value to this country as an indication of our technical skill.[87]

Duncan Sandys, the British minister of aviation, strongly backed the STAC report with the statement, "If we are not in the supersonic aircraft business, then it's really only a matter of time before the whole British aircraft industry packs it in. It's obviously the thing of the future. It may pay. It may not pay, but we cannot afford to stay out. If we miss this generation of aircraft we shall never catch up. We will end up building executive aircraft." [88]

Sandys' recommendation that the British government proceed with the supersonic transport effort was taken, and preliminary design studies were begun. The long-range Mach 1.8 contract was given to the Bristol Aircraft Company and the medium-range Mach 1.2 study to Hawker Siddeley Canada, Incorporated.

In the meantime, the Sud Aviation Company in France had independently decided on a slender delta-wing approach to their new Super-Caravelle, the same general approach favored in the British long-range Mach 1.8 study. Sud Aviation and Bristol designers were on common ground.

After about 2 years of discussions and exchanges of information, the British and French designers came into general agreement on the characteristics an SST should have and on how they would proceed on a joint program. The Supersonic Aircraft Agreement, signed on November 29, 1962, committed both nations to the development of an SST. The agreement contained no break clause — neither country could withdraw from the program without the approval of the other.

61

The absence of this escape clause probably ensured the survival of the Concorde.[89]

The aircraft that evolved from discussions between the British and French designers was essentially an all-aluminum design with a thin ogee-delta wing. It was to be powered by four Bristol/Siddeley (BS–593) Olympus engines in the 35 000-pound thrust category. These straight turbojets were equipped with afterburners for use in takeoff and acceleration to cruise flight conditions. Originally, the aircraft was to carry 128 passengers a distance of 4000 statute miles at a cruise speed of 1450 mph (Mach 2.2). During the course of development, however, the cruise speed was reduced to 1350 mph (Mach 2.05), and the passenger load was reduced to the order of 90 to 100 on transatlantic missions.

Although a number of technical problems were encountered during the development of the Concorde, perhaps the most critical problems were in the political arena. Less than 2 years after the project was launched, the United States announced their program for building a larger, faster SST. BOAC, apparently doubtful that the smaller, more conservative Concorde would have adequate range for the transatlantic operation, placed an order for six U.S. SSTs.[90] Caught by this announcement and the escalating cost of the Concorde — the projected cost had risen to twice the original estimate — the British government opted to get out of the program in late 1964.

The British appeal to France to break the SST agreement was refused, and the French let it be known that there was a binding agreement between the two countries "to develop and produce jointly a civil supersonic transport aircraft." It was clear that the French could well take their grievance to the International Court of Justice at The Hague and could probably win.[91]

The British government did not want to face the embarassment of an international lawsuit while they were trying to get into the European Common Market. Consequently, they assured the French that they would honor the treaty. In reflecting on the decision later, Prime Minister Harold Wilson wrote:

Had we unilaterally denounced the treaty, we were told, we could have been taken to the International Court, where there would have been little doubt that it would have found against us. This would have meant that the French could then have gone ahead with the project no matter what the cost, giving us no benefit from the research or the ultimate product. But the court would almost certainly have ruled that we should be responsible for half the cost. At that time, half the cost was estimated — greatly underestimated as it turns out — at 190 million pounds. This we should have had to pay with nothing to show for it. . . .[92]

The Concorde project survived through several changes in government; it survived the removal of Charles DeGaulle as French president; it survived several strikes in both countries; it survived technical difficulties that required a redesign of the wing and engine exhaust system; and it survived the escalating costs that were to rise from a 1965 estimate of $400 million to a 1977 total cost of about $4 billion,[93] ten times the original estimate.

The first flight of the Concorde was originally scheduled for 1967, and it was to enter passenger service in the middle of 1971. As it turned out, the first test flight did not occur until March 1969, and passenger service did not begin until 1976, with flights from Paris to Rio de Janeiro (via Dakar) by Air France and from London to Bahrain by British Airways. Service from Paris and London to Washington, D.C., began on May 24, 1977. The combined level of service for the two airlines was about 110 flights per month for the first year of operation and rose to about 140 flights per month after the inauguration of flights to New York in December of 1977.[94] In 1982, the level of service was reduced because of a worldwide recession and the escalating price of jet fuel.

The Concorde has been described as a "supersonic bust" [95] and as an aircraft with "disastrous economics." [96] In some respects, these descriptions are accurate. The shortcomings of the Concorde are attributable, in part, to the limited passenger load and to the conservative approach followed in the design. For the first really operable supersonic transport, this design approach was probably in order. The balance of the shortcomings are in the minds of those who are not ready for the supersonic transport. An airplane can hardly be a success if its operators have to fight endless legal and political battles to secure landing rights in American airports that have previously been open or if they must fight a mob that is resisting the plane's approach by lying down on the runway.

The fact remains that, in its first 3 years of operation, the Concorde carried 400 000 passengers over 25 million miles and accumulated nearly 30 000 flying hours.[97] In the next 3 years, these numbers trebled as the Concordes accumulated nearly 100 000 hours of passenger service. This record was achieved without a serious accident and with very low utilization because of limited landing rights.

Although many feel that the Concorde program proved to be economically disastrous,[98] the technical achievement of the Concorde is best summarized by the words of Goeffrey Knight:

> Technically, Concorde's is a triumphant story. Apart from the American space programme, I can think of no other aircraft project involving high technology that has come through with such success. We are probing the frontiers of knowledge all the time, and advanced the state of our art at every stage.[99]

The American SST program was canceled after 8 years without the construction or flight of a prototype. The Soviet Union's SST program is essentially at "parade rest" after the construction and flight of eight to ten aircraft. Sixteen Concordes were built by the British and French before the assembly lines were closed. Only the state-owned airlines of the participating countries purchased the Concorde, and it is the only SST currently in regular passenger service. How long it can maintain this service is yet to be determined.

Why have these three advanced aircraft programs, launched with such high expectations and with the support of their respective governments, failed to really open up the era of supersonic cruise flight? Are there some lessons to be learned from the essential failure of these programs?

CHAPTER 6

Lessons Learned in Pre-1972 Supersonic Cruise Experience

Over the past 35 years, four of the world's most advanced technological nations—Great Britain, France, the Soviet Union, and the United States—have mounted concerted efforts to develop long-range supersonic airplanes. It was once anticipated that these efforts would lead to the wide use of supersonic transport airplanes by the early 1970s. In reality, only the British/French Concorde has achieved a modicum of success, and this success has been tempered by the fact that only a few of these airplanes were built and that no replacement aircraft or advanced supersonic transport is under serious consideration. Although the Concorde is an outstanding technical achievement, its current limited-passenger service certainly does not represent the anticipated coming of the era of supersonic commercial air transportation.

In spite of large expenditures of time and money, the efforts to develop an environmentally acceptable, economically competitive supersonic transport have not been successful. Even with this lack of success, however, perhaps some important lessons can be learned from the efforts. Some of these "lessons learned" are now discussed, not necessarily in order of importance.

THE MILITARY/COMMERCIAL DIFFERENCE

Each of the efforts to develop a successful SST took considerably longer than expected—even the American effort that did not result in a flight vehicle. Also, each of the programs led to substantially higher expenditures than were estimated at the outset. A contributing factor in the underestimation of the time and costs of commercial SST development was the overly optimistic projection of the value of supersonic military flight experience. All of the nations involved in attempts to develop a commercial SST had constructed and flown experimental and military supersonic airplanes. This flight experience had, in fact, provided the data that suggested the feasibility of supersonic commercial flight. And, of course, the engineers involved in the development of SSTs were aware of differences between military and commercial supersonic flight. It is doubtful, however, that the magnitude of these differences was fully understood at the beginning of the SST development.

Clarence L. (Kelly) Johnson, leader of Lockheed's well known "Skunk Works" airplane design team was said to have made the following remark, "Give me a big enough engine and I'll design you an ironing board that'll fly." [100] This statement, whether actually made or not, leans toward a representation of the design philosophy followed in many of the early military supersonic airplanes. Most of these airplanes were designed for maneuverability, high rates of climb, and short-range flight at supersonic speeds. The engines were sized to provide performance rather than efficiency, and little attention was given to noise or other environmental factors. Extended ranges were made possible by aerial refueling or flight at subsonic speeds, recourses that would not be desirable in a commercial SST from safety and economic considerations.

The nearest military prototype for a commercial SST was the American B-70 bomber. In the design of this airplane, efforts were made to achieve the high levels of supersonic cruise efficiency demanded of a commercial SST. However, no attempt was made to meet environmental constraints or economic requirements. The B-70 did not have to operate in the strict environment of a commercial airport, did not have to meet the reserve fuel requirements of an SST, and was not designed for the 50 000 hours of life expected of a commercial transport. In addition, the B-70 operation was not threatened by sonic boom restrictions. Thus, the critical elements in SST design—economics, airport performance, reserve fuel requirements, airplane life, and sonic boom—were not controlling elements in the design of the B-70, nor have they been in any other military design.

With the cancellation of the B-70 program, there was no longer any supersonic military airplane that even remotely resembled the design and operating requirements of an SST. The bulk of the supersonic military airplanes spend only a few minutes of each mission at the searing temperatures experienced at high-speed flight conditions. On the other hand, the SST would be bathed in temperatures of one to two times the boiling point of water for nearly 3 hours on an intercontinental flight. At Mach 2.7 cruise conditions, for example, the nose of the proposed Boeing SST would have been 6 inches longer than when it was sitting on the ground. This heat-expansion problem, not present in most military airplanes, forced Boeing to consider abandoning the traditional primary cable control system in favor of a triply redundant electrical command or fly-by-wire system with cable backup.[101]

The pilot and crew of supersonic military airplanes are provided with pressure suits for protection against sudden decompression at high altitudes. Such protection would not be practical nor acceptable for 200 to 300 passengers on an SST flight. Accordingly, the SST structure has to provide a fail-safe protection against rapid decompression, an event that would bring quick death to the passengers from ebullism—boiled blood—at an altitude of 60 000 feet. At this altitude, there would not be time to reach for the oxygen mask.[102]

In the development of commercial subsonic transports, the step from military to commercial application was small. In some cases, particularly in the era of propeller-driven airplanes, the transition was essentially direct. Approximately 25 000 hours of subsonic jet bomber cruise experience preceded the first American commercial subsonic jet transport design—the Boeing 707.[103] This has not been the case in the supersonic speed regime. The military design criteria for aerodynamic performance, structures, and operations have been so different from the criteria required for an acceptable SST that little technology transfer has been possible, except in the propulsion area.

SUPERSONIC CRUISE TECHNOLOGY

A SUCCESSFUL SST WILL PERMIT LITTLE ROOM FOR DESIGN COMPROMISE

Past experience indicates that there will be little room for design compromises in the development of a successful SST. To meet the stringent environmental constraints of noise, sonic boom, and pollution in a safe, economically competitive SST will require the best possible combination of aerodynamic, structural, and propulsion technologies. Isolated advances in the disciplinary technologies are meaningless unless they can be integrated into a congruent airplane that meets all mission requirements.

Technology integration is important in the development of aircraft for any flight speed. Disciplinary technology advances are not automatically applicable to any airplane until a careful integration study is made, particularly in the case of supersonic airplane design. The performance of all elements of the design are interrelated, and these interrelationships are as important in the overall design as the basic elements. For example, the placement of engines on a supersonic airplane not only affects the performance of the engines, but also affects the aerodynamic and structural design and performance of the airplane. Similarly, the structural design influences aerodynamic performance, and aerodynamic characteristics and flight speed are critical in the structural design and material selection. The miscalculation or simplification of any of these mutual interactions can lead to a failure to meet the overall airplane requirements.

Commenting about the British/French and American SST design approaches, Dr. A. E. Russell of the Bristol Aircraft Company remarked, "The Mach 2 solution offers close competition on similar financial arrangements [to the subsonics] while the Mach 3 hotrod needs a very indulgent backer and an uninhibited operator. . . ." [104] In their decision to develop an aluminum Mach 2 SST with a simplified ogee-delta wing, the British and French perhaps knowingly compromised aerodynamic potential in return for a measure of confidence in structural design and projected costs. In spite of this conservative approach, however, the Concorde development did not go as planned.

68

LESSONS LEARNED IN PRE-1972 EXPERIENCE

Entry into flight service took 50 percent longer than anticipated, and the project ran up costs 10 times the original estimate. Without depreciating the technical achievement embodied in the Concorde, the limited range/payload capability and marginal aerodynamic potential were factors that led to the lukewarm acceptance of the airplane by the world's airline operators.

In further commenting about the American effort to develop an SST somewhat faster than the Concorde, Dr. Russell said, "The advocates of Mach 3 airliners seem to be confronted by a formidable array of self-inflicted difficulty!" [105] Although the final choice of cruise speed for the proposed American SST was only Mach 2.7, the U.S. design was at the opposite extreme from the Concorde in that no element of the design could be considered conservative. The higher temperatures at Mach 2.7 made it necessary to switch from aluminum — a material that had been used on nearly all the previous transport airplanes. The complicated variable-sweep feature of the proposed American SST was chosen in a conscious effort to achieve flexibility of operation and optimum aerodynamic performance, but this feature also posed structural design and weight problems of uncertain magnitude. The Boeing Airplane Company made a concentrated effort over 2 years to reap the potential aerodynamic advantages of variable sweep. As *Fortune* magazine reported:

> Boeing Co.'s attempt to build a supersonic airliner has turned out to be the most bafflingly complicated job of research and development in the history of American industry — and it may end up the costliest. The undertaking has come to demand engineering resources, human and inanimate, second only to those going into the national effort to put a man on the moon. During the five years that Boeing has been engaged in this task on a significant scale — the last two at maximum effort — the company has run up 30,000 hours of research on eight different wind tunnels scattered across the nation, 80,000 hours of analysis on computers that include the largest in existence and a total of 8½ million engineering man-hours altogether — all this to the end of reducing the Boeing vision of the SST engineering drawings from which the shop can begin to fabricate a prototype. [106]

In spite of this effort, the potential aerodynamic advantages of variable sweep could not be reconciled with the structural and weight uncertainties, and the concept was dropped from SST consideration.

SUPERSONIC CRUISE TECHNOLOGY

It should be pointed out that there is nothing wrong with the ogee-delta-wing concept of the Concorde or the variable-sweep feature proposed on the Boeing B–2707 SST. The delta wing has been successfully used on supersonic military airplaines such as the F–102, F–15, and B–58 and on several foreign military aircraft. Similarly, the variable-sweep wing idea is used in the United States on the F–111, F–14, and B–1 airplanes and in the Soviet Union on several military airplanes. However, none of these aircraft are required to meet the operating and economic requirements of the SST. Successful application of a disciplinary concept to one mission did not guarantee its useful application to an SST mission.

The lesson — the importance of technology integration — learned in past SST experience was a principal element in the NASA SCR program that will be discussed later.

U.S. Air Force B–1 variable-sweep bomber made by Rockwell International.

THE VALUE OF FOCUSED TECHNOLOGY EFFORTS

We sometimes like to believe that most of America's advances in the science of high-speed flight have resulted from "basic disciplinary research" conducted in a laboratory atmosphere by dedicated scientists working at the outer fringes of the state of the art. As a matter of fact, this is not the case. American "advanced fundamental basic aeronautical research" provided a totally inadequate foundation for the era of transonic and supersonic flight that exploded upon us in the 1940s. Aeronautical testing methods in the transonic and supersonic speed regimes were crude and inadequate; there were no reliable methods for transonic and supersonic design and analysis; and U.S. technology of turbojet engines was essentially a "buy in" from our British allies. In March 1941, because of reports of German research into reaction powerplants, the Army requested that NACA study jet propulsion. The Army Air Corps chief, General Henry H. "Hap" Arnold, journeyed to Great Britain, where he was surprised to learn that the British were preparing to flight-test a turbojet engine, the Whittle W–1. He subsequently arranged for the development of the Whittle engine in this country by the General Electric Company.[107]

While moving rather slowly in the basic research mode, U.S. technology of high-speed flight made rapid advances under the focus and urgency of America's efforts to "break the sound barrier" and later under the focus of efforts to develop a commercial SST. Within a relatively short time, U.S. engineers and scientists had solved testing problems at transonic speeds, developed methods for use in the design and analysis of high-speed airplanes, improved the facilities for testing at supersonic speeds, developed a fairly complete understanding of sonic boom, and launched a comprehensive technology program for improving the performance of jet engines.

Note also that America's basic space technology was unsuccessful in developing and launching a Vanguard satellite in 1957 during the International Geophysical Year program. Under the leadership of a former German scientist, Dr. Wernher von Braun, and using German technology (Jupiter C modified intermediate range ballistic missile), America's first space satellite, the Explorer, was successfully launched in 1958. After that, with the focus of a national space program, U.S. engineers made rapid and remarkable strides in space technology.

As a result of recognition of the importance and value of focused technology efforts, this approach was followed in the NASA Supersonic Cruise Research (SCR) and Variable Cycle Engine (VCE) programs to be discussed later.

SUPERSONIC CRUISE TECHNOLOGY

U.S. TECHNOLOGY SUBSIDIZATION

A recent report by the Office of Technology Assessment of the U.S. Congress indicated that "One of the most important lessons learned [from the U.S. SST program] is that a genuine and important national interest will have to be clearly identified before any future high-technology large-scale commercial undertaking can expect to receive significant Government support in the future." [108] Although this statement is probably true, it serves to point up anomalies in the subsidization policies of the U.S. government.

One of the fastest growing commercial enterprises in America is the satellite communications industry. This industry has directly benefited from the NASA satellite program and the millions of dollars expended on it. The electric power companies have directly benefited from the billions of dollars that the U.S. government has spent in developing nuclear energy. The U.S. railroads, unable to make it on their own, as well as the mass transit systems and the interstate highway network, have received billions of dollars in government subsidies over the past few years. For years, the U.S. government has provided millions of dollars in subsidies to the maritime interests in this country, with the stipulation that the ships built can be used by the U.S. military in the event of war. Billions of dollars are also spent by the U.S. government each year to purchase surplus commodities from the agricultural community, and additional millions are spent in storage costs for these farm products. In addition, the NASA space shuttle, which is under development by the government at a projected cost approaching $10 billion, is expected to have many commercial uses.

In spite of these obvious subsidies of many commercial ventures, the U.S. government has been reluctant to directly participate in the development of high-technology commercial airplanes. In fact, in the 1940s, the U.S. Congress refused to approve funds for constructing a subsonic jet transport to meet the challenge of the British Comet.[109] If the Comet had not developed problems, this decision could have cost the U.S. aircraft industry the tremendous business it now has in subsonic jet transports. The airplane industry is currently one of the largest employers in the nation and has had a long-term positive effect on America's balance of trade. The air transportation industry, user of airplane industry products, has been a powerful factor in America's rise to economic and technological preeminence. It is difficult to understand the political aversion to direct support of these industries.

Although this aversion was temporarily put aside by President Kennedy and the Congress in 1963 in their support of the U.S. SST program, the uncertainty of government support was one of the factors that was to plague the effort from start to finish.

Government funding was only one of the uncertainties that led to the cancellation of the U.S. SST program. There were worries about environmental impact and passenger safety. There were questions about radiation and air-traffic control. The airlines did not know if the SST would be economically viable or if they could afford such airplanes. The airplane manufacturers had committed heavily to the development of "wide-body" subsonic jet transports and were concerned that the SST would reduce the subsonic market.

In view of these uncertainties, it was difficult to determine if the development of a commercial SST was of genuine and vital importance to the interests of the United States. It was purely a matter of semantics that the subsidization of an SST was different from that of a commercial railroad, a commercial satellite communications company, or a commercial power company. However, Congress certainly could not be expected to continue funding the SST when the people who would build and use it were negative or ambivalent.[110]

EVOLUTION OF AN ACCEPTABLE SST WILL BE DIFFICULT

In the advancement of a technology, it is not generally anticipated that the prototype of an innovation or improvement in the state of the art will provide the ultimate answer in terms of performance. This ultimate answer will come after evolutionary refinements and improvements in the finished product. However, the prototype is expected to demonstrate performance and provide a firm basis for future improvements in both performance and costs. This process has been followed in the development of the air transportation industry. The Boeing 707 and the Douglas DC–8, prototypes of the highly efficient subsonic jet transport fleet, represented a revolutionary step in air transportation. From these original prototypes, more efficient, more productive, more comfortable, and safer subsonic jet transports have evolved.

The experiences of the Concorde and the U.S. SST programs indicate that the evolution of an acceptable SST will be difficult. The Concorde represents a technological triumph for its British and French developers. However, in the eyes of its potential customers—the airlines—the Concorde did not represent a suitable prototype for development into a family of advanced SSTs. Consequently, the Concorde has not been allowed the evolutionary cycles that were so important in the development of subsonic transport technology. In the case of the U.S. SST, enough uncertainty existed about the acceptability of the design to preclude further funding of the project.

Almost no one will deny that "the SST presents the greatest challenge in aviation history. . . ."[111] Consequently, perhaps too much was expected of both the Concorde and the American SST program. Both programs were conducted during a period of social unrest and upheaval, and both efforts sought to bring about a quantum jump in the technology of air transportation. Both efforts anticipated that the unacceptable prototypes would evolve into acceptable SSTs through future advances in technology. Because of the nebulous nature of these projected technology advances and the mounting costs of the projects, however, the evolutionary phases did not come to pass.

In any future effort to introduce the SST into the air transportation system, it appears that the prototype SST must be acceptable in nearly every respect. "There will be no room for unsuspected bugs in the SST." There will also be no margin for error, and this is the greatest challenge of all, for no commercial airliner ever built has been completely free from design bugs.[112] The available technology must be more than adequate to sustain the program, and answers to all the

74

uncertainties must be in hand. It is unlikely that an acceptable SST will be permitted the luxury of evolving from an unsatisfactory prototype base.

PERCEPTIONS OF A "RAMPANT" TECHNOLOGY

It is quite possible that the Concorde and the American SST would be in wide use today had it not been for a rising sentiment that technology had "gone wild." Citizens of the highly developed nations were becoming more and more concerned about noise, pollution of rivers and the atmosphere, and what they perceived to be a misuse of our natural resources. Unchecked technological growth was seen to have been responsible for slums, pollution, and even the Vietnam War.[113] The SST became a symbol of technical arrogance and an object on which anti-technology sentiment could be focused.

Many people who had gratefully acknowledged and taken advantage of great advances in air transportation felt that these advances had gone far enough. The horror stories about SST noise, sonic boom, fuel use, and atmospheric pollution were enough to convince large segments of the population that the SST was a harbinger of catastrophe. Anti-SST groups, which began to form in the United States and Europe, initiated concerted efforts to stop the Concorde and the U.S. SST programs.

The organized efforts against the SST were perhaps the first evidence of a powerful new area of politics — the politics of technology. Fundamental to the new politics was the choice that industrial society made about what priority it allocated to technical growth and what priority it gave to the environment. At stake were such important factors as the public's right to question governmental decisions in areas that might affect drastically the quality of life for everyone.[114]

The politics of technology, which surfaced while the American SST was under development and at about the time the Concorde flight tests began, became a factor in these and other technology programs. Efforts to develop nuclear energy, to make use of oil shale and coal deposits, to

75

construct oil and gas pipelines, and to search for new mineral deposits came into the sphere of the new political force.

The opponents of the SST had legitimate questions and concerns about what the operation of such a vehicle would do to the environment. They also had legitimate questions as to why the SST was being developed and why the governments were paying for it. Unfortunately, SST proponents waited too long to face up to these questions. In the meantime, the anti-SST forces were providing their own answers, which in many cases are now recognized as exaggerations or half-truths. Consequently, the SST was unable to assume its legitimate role as the next revolution in the air transportation industry. Rather, it became billed as a development program for the benefit of "jet-set playboys and their ladies" [115] and one of the most resented technological programs of all time.

It cannot be stated with certainty that the "politics of technology" led to the cancellation of the U.S. SST program or that it has been a factor in the reluctant acceptance of the Concorde. However, it is almost assuredly true that any future SST effort in this country will be subject to this new political force.

SUMMARY OF LESSONS LEARNED

In summary, some of the lessons learned from the pre-1972 supersonic cruise experience are as follows:

- Major differences exist between the technologies required for a supersonic military aircraft and an acceptable SST. Aeronautical technology that is applicable to supersonic military aircraft may be totally useless for a commercial SST.

- A successful SST will require the best possible combination of disciplinary technologies. An exciting technical innovation in one discipline is unusable if it cannot be successfully integrated with the other disciplines.

- Focused technology efforts will bring about a more rapid improvement in the aeronautical state of the art than will basic generalized research.

76

LESSONS LEARNED IN PRE-1972 EXPERIENCE

- U.S. subsidization policies are rife with anomalies. A stronger case must be made for the SST or any other high-technology item before much government support can be expected.

- A future evolution of an acceptable SST is not likely. The technology for a satisfactory prototype must be in hand, and answers to the uncertainties must be available.

- The politics of technology will be a powerful factor in any future high-technology effort.

These "lessons learned" had an impact on the formulation of the NASA post-1972 program in supersonic technology. But, why was there a post-1972 NASA supersonic technology program? What were the problems to be worked on? After the SST program was canceled, is there still some promise in the concept of supersonic cruise flight?

CHAPTER 7

Supersonic Cruise Problems and Potential

Senator William Proxmire, an astute politician and outspoken critic of U.S. government funding of the SST, said in 1970 that "it seems utter folly to continue with the production of this plane (the SST) until we have developed the technology necessary to make supersonic flight compatible with respect for our environment." [116] At the same time, Representative Henry Reuss said, "Considering increased airport noise, sonic booms, air pollution, and potentially harmful weather changes, the SST, for which the American public is being compelled to pay, is an environmental outrage." [117] On the other hand, Representative John McFall commented that, "It seems to me that the SST has become a symbol of the need to improve our environment, and in a sense it is a false symbol because the SST is really not that important to the environment, but it has become sort of a rallying point for those who want to improve the environment." [118]

The environmental argument certainly was a factor in the cancellation of the U.S. SST program and the negative acceptance of the Concorde. The deeper underlying reason for the failure of the SST, however, was the fact that the level of aeronautic technology available in the 1960s was not adequate to sustain the development of an economical SST. It was expected that, when the SST was under way, solutions could be found to the critical technical problems. In the end, the SST would meet everyone's concept of what it should be.

The expectations of the SST proponents and developers were neither unreasonable nor unusual. Many high-technology programs had

79

begun with less than all the technical answers to become successful additions to the state of the art. However, the lengthy, expensive, and sometimes fitful efforts to develop supersonic technology during the course of the Concorde and American programs provided arguments for SST opponents and detractors. These efforts also served to alienate the interest of SST supporters, the people, the governments, and even the developers. What are some of these critical technical problems of supersonic cruise airplanes?

TECHNICAL PROBLEMS

AERODYNAMIC PROBLEMS

The level of aerodynamic efficiency (lift/drag ratio) at supersonic cruise speeds is a critical factor in the performance of a supersonic cruise airplane. All other things being equal, the higher the supersonic lift/drag ratio, the greater the range potential for a given fuel supply or the lesser the fuel requirement for a given range. If no other factors were involved, the design with the highest value of supersonic lift/drag ratio would lead to the most efficient supersonic cruise airplane. According to all available aerodynamic theories and tests, a supersonic design utlizing a highly swept arrow-head wing would provide the highest attainable levels of supersonic lift/drag ratio.

As a matter of fact, the supersonic cruise lift/drag is not the only critical aerodynamic factor in the design of a supersonic cruise airplane, particularly a commercial SST. An SST is also expected, and required, to operate efficiently at subsonic speeds. The reserve fuel requirements — the fuel required for flight to an alternate destination and holding for landing clearance — are based mainly on subsonic lift/drag ratio. The noise characteristics of the airplane are also related to low-speed aerodynamic performance. For these low-speed flight conditions, theory and experience suggest that a moderately swept-wing design provides the highest levels of aerodynamic efficiency. Highly swept wings were seen to provide relatively poor aerodynamic performance in this low-speed regime.

PROBLEMS AND POTENTIAL

As a possible means of resolving the conflicting wing-sweep requirements in the subsonic and supersonic flight modes of a supersonic airplane, NASA proposed the use of variable sweep — a concept that involves altering wing sweep during flight to the most optimum position. This concept has been successfully adapted in the design of military aircraft, in which range/payload performance and efficiency are not critical design factors. However, during the SST program, it proved to be unfeasible because of weight and integration problems. Both the Concorde and the American SST developers finally chose a swept-delta or delta-ogee wing design that was a compromise between subsonic and supersonic lift/drag ratios. Although understandable, this compromise was a limiting factor on the growth potential, range/payload characteristics, and economic feasibility of the two airplanes. Technology was not available to permit the achievement of the necessary optimum or near-optimum aerodynamic efficiency at supersonic cruise conditions while simultaneously meeting the other design and operating re-

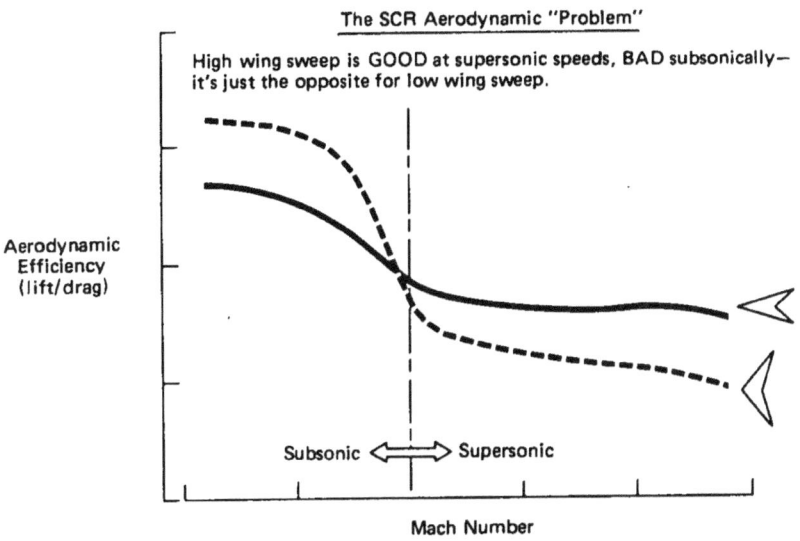

Supersonic cruise aircraft aerodynamic "problem."

quirements. The development of this high lift/drag (arrow-wing) technology could dramatically improve SST performance.

PROPULSION SYSTEM PROBLEMS.

In developing a propulsion system for supersonic cruise airplanes, the engine designer runs into the same incompatible flight conditions faced by the aerodynamicist. At supersonic speeds, engine thrust or power is provided most efficiently by moving a relatively small volume of air at high velocities, a characteristic of the straight turbojet engine. At subsonic speeds, however, it is more efficient, and quieter, to move a larger volume of air at fairly low velocities, a feature of the turbofan engine. For application to an SST, a turbojet engine sized to give optimum performance at supersonic speeds does not provide adequate thrust at takeoff and during climb conditions. The turbojet also requires more fuel for subsonic overland operations and for alternate

Problem 1.: The turbojet engine is more efficient than high-bypass turbofan at supersonic cruise conditions, but high-bypass turbofan is more efficient for subsonic operations.

Problem 2.: High jet-velocity turbojet engine is desirable for supersonic cruise but is noisy around the airport. The inverted velocity profile (IVP) coannular nozzle illustrated in inset is a characteristic of VCEs and gives inherent noise relief.

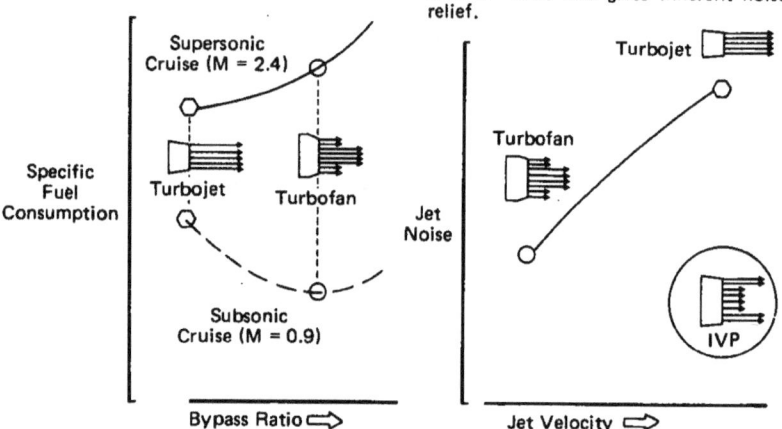

Supersonic cruise aircraft propulsion "problem."

field and "hold" reserves than the turbofan. On the other hand, a turbofan engine would be quieter than a turbojet engine, but would require more fuel for the critical supersonic cruise condition. The turbofan engine also requires a larger diameter, with increased drag, than the turbojet to provide the same thrust level.

Turbojet engines were selected for both the Concorde and the U.S. SST because of their more efficient performance at the critical supersonic cruise conditions. Afterburners were incorporated to augment the basic engine thrust for takeoff, climb, and acceleration to cruise speed and altitude. The afterburner essentially increased the basic jet exit velocity and, hence, thrust by burning fuel in a duct at the rear of the engine. The afterburner proved to be a noisy way to produce thrust and made it unlikely that either the Concorde or the U.S. SST could meet the new Federal Aviation Regulation (FAR), part 36, noise rule enacted by the FAA in 1969. In the latter stages of the American program, an oversized nonafterburning turbojet engine was considered in an effort to meet this new noise rule. The weight and size of the engine had a catastrophic effect on the American design. In the belief that the FAR 36 noise rule would not be applied to an airplane already under development, the Concorde retained the afterburning turbojet engine.

The afterburning turbojet engine has been the major powerplant for supersonic military airplanes for which engine noise and subsonic flight efficiency are not primary design parameters. Large turbofan engines are now used almost exclusively on subsonic jet transports because of their superior noise characteristics and subsonic fuel efficiency. Because a commercial SST must operate efficiently at both subsonic and supersonic speeds, it could use an engine that incorporates the best features of the turbojet and turbofan engines. This technology was not available during the SST program, and the turbojet engine with afterburner was selected as the best compromise.

Other technical problems in the propulsion discipline impacted the development of an SST. For example, the inlet is as critical to engine performance as the engine itself. Unless the inlet can deliver air to the engine in an efficient manner, the engine cannot perform efficiently on an overall basis. The nozzle, or engine exit, also contributes importantly to propulsion system performance and noise. Both the inlet and nozzle, as well as the engine itself, require a high degree of control for maximum performance. Although considerable work was done on inlet and nozzle configurations during the SST program, further intense effort was required, particularly with a new engine cycle. Questions also remained about engine emissions and pollution, particularly on afterburning engines. This problem became a critical issue in the latter stages of the U.S. SST program, when fear was expressed that nitric oxide (NO_x) emissions from SST engines would seriously deplete the protective ozone layer in the atmosphere and lead to a substantial increase in skin cancer. The technology was not available to allay these fears or to effectively reduce the emission levels.

STRUCTURES AND MATERIALS PROBLEMS

Structures and materials technology for building supersonic airplanes was certainly available in the 1960s. A number of supersonic military airplanes were already in service, and an advanced titanium supersonic reconnaissance airplane, the SR-71, was under development. As the SST program proceeded, it became readily apparent that the structural design of an SST presents different challenges than the design of a military airplane and has more impact on the acceptability of performance. Structural materials are also used differently on an SST than on a military aircraft.

The structure of a military airplane is usually rigid because of the requirements of maneuverability. The weight, safety, and cost of the structure are important, but have no direct bearing on whether the aircraft is considered satisfactory for its design mission. On the other hand, the structure of an SST is generally quite flexible because it has no harsh maneuver requirement and because the structure is usually lightly loaded. The weight, safety, and cost of the structure are extremely important factors in the acceptability and economic performance of an SST.

Because of the flexible nature of SST structure, the relationship between structural and aerodynamic design is very important. The differing aerodynamic forces during the takeoff, subsonic cruise, climb and acceleration, and supersonic phases of flight alter the shape of the flexible structure. The shape of the structure in turn influences the aerodynamic forces. An accurate assessment of these mutual interactions is necessary for achieving an optimum overall design.

During the course of the SST program, supersonic structural design methods were available. These methods were cumbersome and time consuming, however. It was not unusual to make a configuration change that was expected to improve performance, only to find that an adverse weight increase due to the configuration change had canceled the anticipated improvement.

Another structures problem exposed by the SST effort was the absence of well-defined, low-cost fabrication for supersonic high-strength aluminum and titanium structures. The mounting costs of the development phases of the Concorde and the U.S. SST were significantly affected by fabrication cost, and the projected production costs made the SST a questionable economic venture.

The U.S. decision to develop a Mach 2.7 SST posed a materials problem that was not present in the Concorde program. At the higher temperatures experienced at Mach 2.7 flight conditions, the available technology for necessary nonmetallic materials such as fuel-tank sealants was inadequate. This problem had not been resolved at the time the American SST program was canceled.

TECHNOLOGY INTEGRATION PROBLEMS

The pre-1972 supersonic cruise experience indicated the overwhelming importance of technology integration in developing a successful SST. Perhaps no other high-technology program besides the SST has been considered unsuccessful when its performance goals have been met. The Concorde essentially met its design expectations in aerodynamic performance, structural weight, propulsion system efficiency, safety, noise, sonic boom, etc. However, many consider the airplane to be a failure because the levels of technology, although successfully employed, were not adequate to make the Concorde an economical transportation system.

The American SST program was unsuccessful because the advanced levels of technology employed were never integrated into a coherent airplane. The complicated variable-sweep and highly swept arrow-wing concepts offered dazzling advantages in aerodynamic perform-ance, but the structural design methods and materials were not available to make them work within acceptable weight limitations. The relatively simple structural concepts of the delta wing offered fantastic weight advantages, but had insufficient aerodynamic potential to promise more than a marginal SST. Propulsion systems that offered optimum performance at supersonic speeds were not very good at sub-sonic speeds, and vice versa.

The experiences of both SST programs indicate that the develop-ment of a successful supersonic transport will depend more on the ad-vancement of the state of the art of integrated technology than on the advancements in the separate disciplinary technologies. In the least, the disciplinary technologies must be developed within the focus of the integrated airplane and mission requirements.

ENVIRONMENTAL PROBLEMS

The pre-1972 supersonic cruise experience exposed a number of en-vironmental problems that were either critical to the acceptability of the SST or unanswered by its proponents. The major environmental issues associated with the SST were:

- Engine noise
- Sonic boom
- Engine emissions
- Radiation exposure.

Although only two of these issues—engine noise and sonic boom—proved to be significant, all of the environmental questions merit consideration and will be factors in any future SST effort. Some people consider the anxiety and uncertainty caused by these environmental issues to be the factors that finally led to the cancellation of the American SST effort.

ENGINE NOISE

Any piece of machinery that produces work, power, or any other useful product creates noise. The SST engines would certainly be no exception. Each of the four Concorde engines produce on the order of 35 000 to 40 000 pounds of thrust at takeoff, and the four engines on the U.S. SST would have produced about 60 000 pounds of thrust each. Both the British/French and U.S. engines made use of afterburners, and both engines were developed before the promulgation of the FAR 36 noise rules in 1969. Neither engine could meet the noise rules, but it was generally believed that the noise rules would be waived since both SST programs had been initiated before the rules were adopted.

In the early 1970s, the support for the American SST was waning, and the Concorde was undergoing flight tests. At this time, public sentiment, driven by widely held perceptions that all jet aircraft are noisy, began to swing heavily against the noise that was to come from the SST. In an effort to win support for the flagging American effort, a belated attempt was made to bring the proposed U.S. SST within the 1969 noise rules. To do this would require the development of a new engine, a further delay in the program, and substantially more money. This gambit did not work, and the Concorde developers were left to fight the SST noise battles.

By the time the U.S. SST program was canceled in 1971, it was too late for the Concorde to meet the new noise rules. The airplane was under construction, and it proposed to fly in the face of the rising public clamor over SST noise. The Concorde developers maintained that the Concorde noise would be no worse than that of the early subsonic jet transports, which were still in use throughout the United States, Europe, and the rest of the world. Although the Concorde essentially produced the same levels of noise as the early transports, the

87

stance of the Concorde developers led to a showdown with the American public in 1977 when the Concorde sought landing rights in the United States. After a great furor, the Concorde was granted a 16-month trial period to operate into and out of the Dulles Airport in Washington, D.C., and, later, to John F. Kennedy Airport in New York City. Since the hectic public demonstrations died down after the first few Concorde flights, little has been said about Concorde noise. Because of its unique operating capabilities to climb and turn away from the heavily populated regions, the Concorde has added very little to the noise exposure of the citizens of Washington, D.C., or New York City.

Even though the SST has not been the noise "monster" that many people expected it to be, noise is one of the major technical problems to be solved before any future generation of SSTs will be acceptable. An aircraft engine, or any other machine, cannot be expected to produce 60 000 to 70 000 pounds of useful force without producing noise. However, it is possible to reduce the noise levels substantially below those of the first generation SST without compromising the ability to meet other SST requirements.

A NEW SOUND BARRIER—THE SONIC BOOM

When Chuck Yeager broke the so-called "sound barrier" with his flight in 1947, he introduced the world to an even more intractable barrier to acceptable supersonic flight—the sonic boom. This nerve-shattering disturbance created a ground swell of resentment toward supersonic flight and represents one of the most difficult aeronautical problems to be faced by the technical community.

The sonic boom is essentially a direct result of supersonic flight. Disturbances in air can travel only at the speed of sound and, hence, cannot move ahead of an aircraft that is traveling at supersonic speeds. Consequently, a sharp pressure pulse forms and is swept behind the airplane to form a conical surface in which the pressure and temperature are locally higher than in the surrounding air. This conical surface follows the supersonic aircraft along its flight path. When a point on this conical surface passes over an observer on the ground,

PROBLEMS AND POTENTIAL

The sonic boom pressure disturbance, Δp, is a constant companion of an airplane in supersonic flight and was one of the major environmental concerns of the SST program. The furor caused by this phenomenon led to the passage of an operating rule that forbade commercial supersonic flights over the continental limits of the United States. As indicated by the inset sketch, the large size of the SST permits it to retain the "near-field" characteristics at the ground. For many flight conditions, the SST sonic boom levels can be reduced by shaping the airplane.

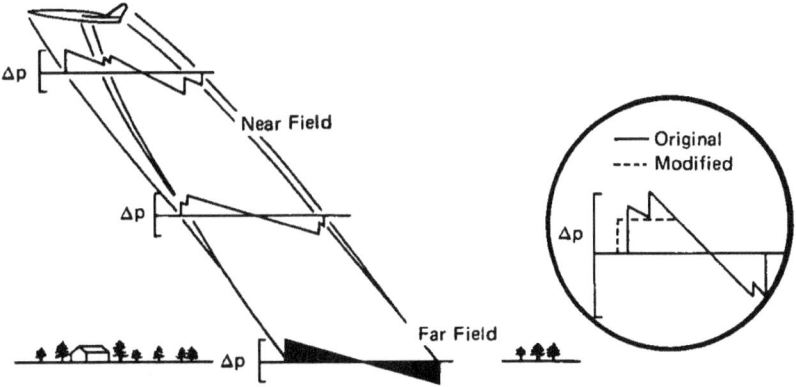

Sonic boom disturbance.

there is a rapid increase in pressure, which the observer perceives as a boom.[119] The level of sonic boom expected of SSTs was 2 to 3 psf, which represents a rather miniscule change in the ambient pressure at sea level of 2116 psf. The abrupt nature of the pressure disturbance startles the observer, and this is the factor that makes the sonic boom such an environmental nuisance. The sonic boom levels created by an SST are not expected to cause any appreciable property damage.[120]

In the early phases of the U.S. SST program, the permissable sonic boom levels for an SST were set at 2 psf in the climb and acceleration phase of flight and 1.5 psf at supersonic cruise conditions. These restrictions severely affected the ability of the SST evaluation airplanes to make a reasonable range/payload mission. Consequently, the sonic boom restrictions were relaxed to a level of 2.5 psf during climb and acceleration and 1.7 psf at cruise. The public fears over sonic boom continued, however, and some people expected the SST program to be canceled because of sonic boom considerations alone. To allay these fears, and perhaps to save the program, Congress passed a law in 1971 prohibiting supersonic flights of civil aircraft over the continental United States. This law, which is still in effect, made the potential American SST a strictly over-water airplane with a more restricted market.

The sonic boom is an irreversible fact of supersonic flight.[121] Means have been found to reduce the levels of sonic boom caused by a large supersonic aircraft and to alter the shape of the disturbance.[122] However, no realistic ways have been discovered to eliminate the disturbance completely. This problem remains a challenging technical goal.

ENGINE EMISSIONS

In the early 1970s, shortly before the SST program was canceled, there was concern that the engine emissions from a fleet of supersonic transports would have a drastic adverse effect on the chemistry of the upper atmosphere. The greatest of these fears was that the nitrous oxide emissions would deplete the ozone in the atmosphere, reduce the shielding from the Sun's ultraviolet rays, and thus cause an increase in the incidence of skin cancer. This concern, originally directed at the anticipated supersonic aircraft, spread to the potential impact of the growing fleet of subsonic aircraft.[123]

As mentioned in Chapter 4, these concerns led the U.S. Congress to instruct the Department of Transportation (DOT) to conduct a scientific study that would provide information for assessing the potential ozone-depletion effects. This major study, the Climatic Impact Assessment Program (CIAP), drew on nine other Federal departments and agencies, seven foreign agencies, and the individual talents of 1000 investigators. The results of this study were not conclusive because of the simplified atmospheric models that were available. Indications were, however, that the first generation of SSTs would cause climatic effects that are much smaller than those minimally detectable. The report went on to say that if high-flying aircraft (including subsonic aircraft) increase greatly in number beyond the year 1980, improvements over 1974 propulsion technology will be necessary to ensure that emissions do not significantly disturb the stratospheric environment.[124]

A subsequent High Altitude Pollution Program (HAPP), conducted by the FAA, indicated that the earlier DOT CIAP study had substantially exaggerated the extent to which future aircraft will reduce the ozone layer. Present understanding of the phenomena indicates much smaller impacts and perhaps no net impact at all.[125] Encouraging as these results may be, the question of atmospheric pollution by air transport engines will continue to be raised. The technical challenge is to continue to search for and find means for reducing the levels of undesirable engine emissions.

RADIATION EXPOSURE

Because supersonic transports will cruise at higher altitudes than previous commercial aircraft, there will be less atmosphere to filter out radiation from outer space. This opens the possibility that the crew of an SST might undergo excessive radiation exposure. It may be rationalized that the SST crew will be exposed to more intense radiation over shorter periods because of the higher speed and reduced trip time for a given flight. This factor would tend to compensate and equalize the total radiation exposure of supersonic and subsonic crews. The best evidence to date is that such radiation exposure will not exceed permitted occupational levels.[126]

OTHER SUPERSONIC CRUISE PROBLEMS

Several other problems were evidenced by the prior supersonic cruise experience, such as fuel usage, cost, economics, and air-traffic control. These factors are almost directly related to the status of technology and the severity of the environmental restraints. These issues will have to be considered in any future supersonic transport program.

SUPERSONIC CRUISE TECHNOLOGY

SUPERSONIC CRUISE POTENTIAL

As technical problems and environmental issues eroded support for the U.S. SST program in the early 1970s, the real reasons and potential for supersonic cruise flight were obscured in the conflict that eventually led to program cancellation. Many of the arguments that had been used to sustain the program either were proven false or were much less important than had previously been claimed. Other supportive arguments were difficult for the average person to understand and, hence, were not helpful in bringing public support to the program.

Among the arguments used to support the development of a U.S. SST were:

- National prestige/foreign competition
- Technological fallout
- Employment
- Balance of trade
- Increased airline productivity.

NATIONAL PRESTIGE/FOREIGN COMPETITION

Both the NASA and FAA recommendations for a national SST program cited prestige and the threat of foreign competition as major reasons for initiating such a program. At the time these arguments were used, they were probably valid. If either or both of the foreign SST efforts were to result in a viable supersonic transport, it was in the interest of the United States to do likewise. However, it is almost axiomatic that prestige and foreign competition are not normally sufficient motivation for a commercial venture. This sort of venture is usually made to capture and sustain a market for the product and to make a profit. Certainly, the British/French consortium and the Soviet Union introduced their programs in an effort to recapture some of the air transportation market that they had lost to the United States.

92

PROBLEMS AND POTENTIAL

When it began to appear that neither the British/French nor the Soviet SST programs would be economically successful, the arguments about national prestige and foreign competition were no longer able to generate support for the U.S. SST program. Even so, the answers to this argument are probably not all in as yet. By carrying their respective SST programs to flight hardware, the foreign nations have gained a mass of supersonic experience that is not available to the United States.

TECHNOLOGICAL FALLOUT

A major argument of SST proponents was that an American SST effort would provide a technological fallout that would be valuable to the aircraft industry in general and to other industrial and military applications.[127] This had certainly been the case in other high-technology ventures.

As mentioned in previous chapters, the focus of the SST program provided major improvements in the state of the art in supersonic design and analysis, sonic boom, supersonic propulsion, structures and materials, advanced navigation systems, etc. Completion of the SST program to prototype flights would probably have brought further technological gains. However, if the major goal — a viable SST — is not attained, the technology fallout would probably not be worth the expense of the program. A costly program cannot be sustained on the basis of technology fallout alone.

EMPLOYMENT

Proponents of the SST cited the favorable effect on employment as a principal reason for continuing the SST program. It was claimed that the U.S. SST program would provide a direct labor force of 50 000 highly skilled jobs with potential application throughout the United States. Taking into account the multiplier factor, the SST program could have more reasonably affected 150 000 jobs.[128]

Although everyone was in favor of employment, it was easy to say that these workers could be applied to programs of more value than that of the SST. On a few occasions, the United States has instituted programs for the major purpose of providing employment. One of these was the Work Projects Administration (WPA) during the depression of the early 1930s. The U.S. government has also awarded high-technology programs to companies in regions of high unemployment. However, no expensive high-technology effort has been either started or maintained on the basis of the employment it would provide.

BALANCE OF TRADE

The U.S. balance-of-trade argument for the SST was one of the most difficult arguments for the public to understand. Figures were quoted on U.S. balance of trade both with and without an American SST, with and without an advanced Concorde, with and without a successful Soviet TU–144, and nearly every combination in between. The impacts ranged from a positive U.S. trade balance of $16.6 billion if the United States would compete with an advanced Concorde to a negative U.S. trade balance of $18.7 billion if we would not compete with the Concorde, a total swing of $35.3 billion.[129]

These balance-of-trade figures became moot when the Concorde program proved to be less than successful, and no plans for the development of an advanced Concorde became apparent. Many people assumed, with relief, that the relative economic failure of the Concorde would remove the need for any further consideration of the influence of the SST on U.S. trade balance. This assumption represented a subtle change in the U.S. reaction to foreign competition. When the British Comet subsonic jet program faltered in the 1950s, the American aircraft industry moved in with subsonic jet transports, which subsequently captured the largest part of the subsonic jet transport market. Foreign sales of these early jet transports and derivatives have consistently provided the most favorable positive effect on the U.S. balance of trade. On the other hand, when the Concorde program faltered, many in the United States dropped the SST and went on to other pursuits. It is difficult to assess what would have happened if the U.S. airplane industry would have been ready to move in with a superior SST.

PROBLEMS AND POTENTIAL

It is also difficult to make a meaningful balance-of-trade argument based on assumptions of any number of predicted events happening, estimating a number of future markets for airplanes of unknown characteristics, etc. One thing is certain, however. The U.S. airplane industry is one of our greatest assets and has been one of the major positive factors in the foreign balance of trade. It is to the advantage of the U.S. government and the American people that this industry remain competitive.

INCREASED AIRLINE PRODUCTIVITY

A recent report published by the Office of Technology Assessment (OTA) of the U.S. Congress stated as one of its findings that "The most compelling argument for an advanced supersonic transport is improved productivity — seat-miles generated by an aircraft per unit time." [130] Other than the SST advantages of improved comfort and reduced trip time, the OTA statement might be carried further to say that there is no real reason for an advanced SST unless it *provides increased productivity within desired economic and environmental constraints.* This is the true goal of supersonic cruise flight and the challenge of supersonic technology.

Over the past 50 years, the demand for commercial air transportation has increased at a phenomenal rate. In the United States alone, the demand grew from 93 million revenue-passenger-miles (rpm) in 1930 to nearly 163 billion rpm in 1975. This represents a remarkable 1750-fold increase in air traffic during the period. Perhaps the most astonishing aspect of the air transportation growth picture is that only a 4.5-fold increase in the number of transport airplanes has been required to meet this 1750-fold increase in air travel demand. Until now, the U.S. airplane and propulsion companies, unfettered by the size, speed, and utilization limitations that have led to the decline of the ship, train, and motorbus as principal intercity carriers, have met this increasing travel demand by providing successive generations of more productive air transports. The continuous evolution of these advanced,

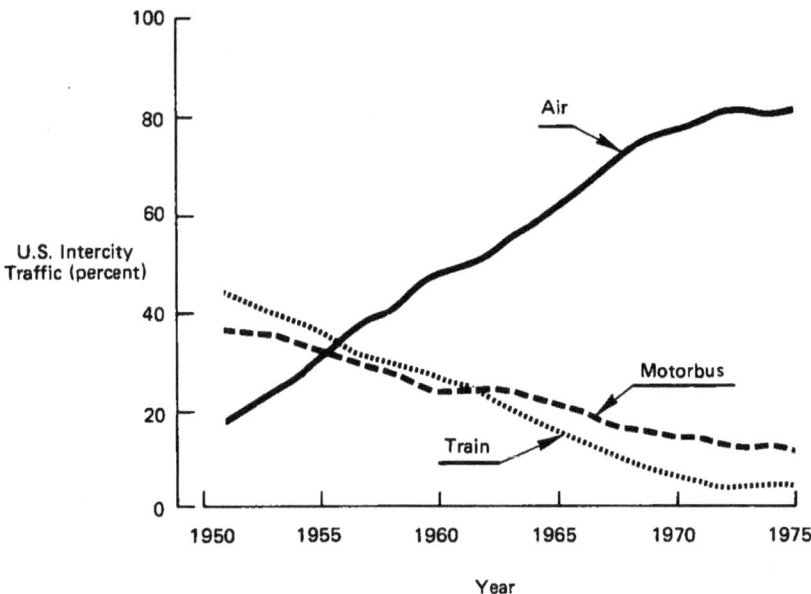

SUPERSONIC CRUISE TECHNOLOGY

As a result of productivity and comfort, the airplane is taking over the transportation market (source: Air–CAB Bureau of Accounts and Statistics; Train/Motorbus — Interstate Commerce Commission).

more productive air transports has successfully prevented air transportation from entering the period of stagnation and decline experienced by other commercial transportation systems.

Productivity has become a widely used yardstick for measuring progress. The Gross National Product (GNP), as well as the rate of growth of the GNP, of the United States, for example, is closely followed by economists. The continuous aim of the U.S. industry is to increase the amount of a given product that is created per unit cost. There is no particular gain to the industry, or the country, if twice the manpower, twice the time, or twice the cost is required to generate twice the product.

The product of any transportation system is revenue-passenger-miles, and the major measurement of the progress of the.system is the improvement in product per unit of time or per unit of equipment. For an individual transportation unit or for the entire transportation fleet, the product can be written:

PRODUCT (rpm) = *Revenue Passengers* × *Speed (mph)* × *Hours*

The first two terms on the right-hand side of this equation (revenue passenger miles per hour) is considered to be the productivity of the given unit or fleet.

In the foregoing equation, the *Revenue Passengers* term is related to the size of the transportation unit or fleet, since the larger the unit or fleet the greater the number of revenue passengers that can be transported. The *Speed* term is simply the velocity at which the revenue passengers are transported, and the *Hours* term is the number of hours that the unit or fleet is operated during the period that the *PRODUCT* is to be measured. The fleet total hours can be increased by improving the utilization of a given unit or by providing more units. Thus, we can say that the principal elements of the *PRODUCT* of a transportation system are *size, speed,* and *utilization hours.*

In the past, the U.S. aircraft industry has met the need for improvements in fleet product or productivity by making simultaneous increases in size, speed, and utilization hours with each successive generation of transport aircraft. Of course, the fleet productivity did not immediately reach the productivity levels of the new aircraft generation because older aircraft remained in the fleet. Gradually, however, the impact of the new aircraft was felt, and the average fleet productivity improved. During the 25-year period 1951–1975, this improvement in average fleet productivity led to the emergence of the airline as the major intercity passenger vehicle in the United States with over 80 percent of the traffic. During this same 25-year period, the Civil Aeronautics Board statistics were of sufficient detail to determine the relative contributions of increased size, speed, and hours to the tremendous increases in U.S. airline product. Each element made an important contribution.

It should be recognized that improved airplane productivity, whether achieved through increased size, speed, or utilization, will per-

SUPERSONIC CRUISE TECHNOLOGY

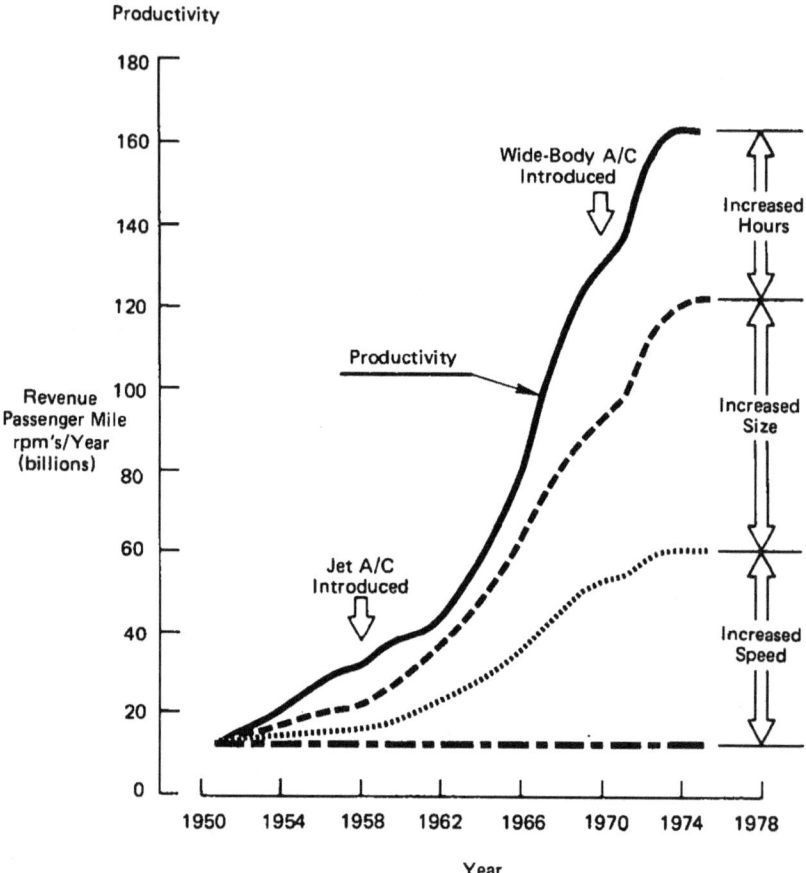

The elements of the airplane's increase in productivity — increased speed, size, and hours (source: CAB Bureau of Accounts and Statistics).

mit a reduced number of aircraft to carry a given passenger traffic. The high productivity of an advanced SST, for example, could limit production to 200 to 400 aircraft and would, of course, affect initial cost and break-even point of the Return on Investment. However, reduced number of aircraft to do a given job is one of the principal aspects of increased productivity in the air transport business.

98

PROBLEMS AND POTENTIAL

If past trends continue, the demand for air travel will increase significantly in the future. To meet this demand, it will be necessary to increase the size, speed, or number of air transports or to improve the use of the airplanes that are currently in the airline fleet. The option of increased size offers some room for improvement, but an airplane can get too big to be effective on anything but a specific route. Similarly, small improvements can be made in aircraft utilization, but there is a limit to the number of hours in a day that passengers want to depart from or arrive at an airport. Therefore, the only options for meeting further increases in air travel demand are to increase the number of air transports in the fleet and/or to increase the speed. Increasing the number of air transports in the fleet will meet passenger demand but will soon result in congestion in the terminal or congestion in the air. It

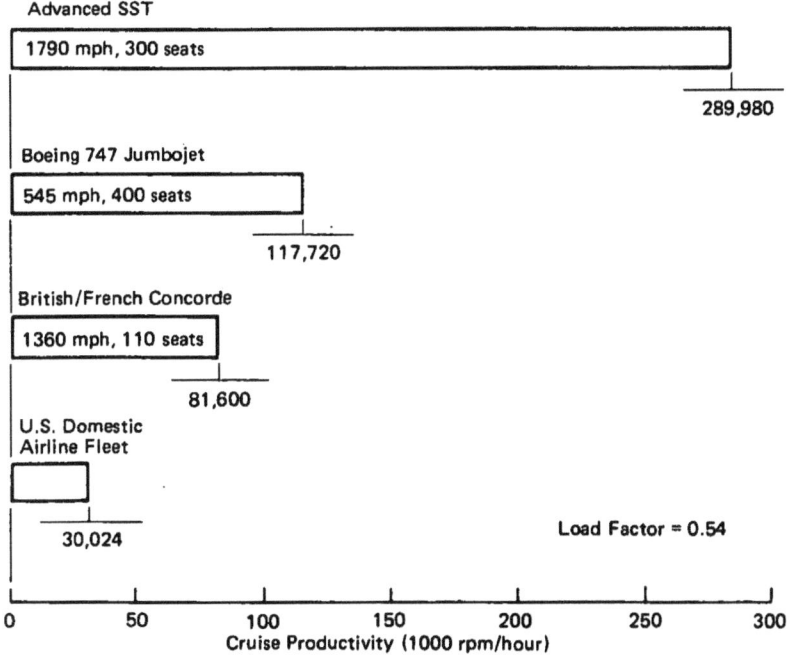

Productivity levels of aircraft (plus for SST).

will not increase fleet productivity. Consequently, the only real hope for meeting the increased demand for air transportation in the future is to again remove the barriers to increased speed that Capt. Chuck Yeager removed some 35 years ago.

The goal and promise of supersonic cruise technology is to again open the air transportation industry to the advantages of increased speed — not speed for speed's sake, not speed for the jet-setter's sake, but speed for the reason that it is a necessary ingredient of productivity. Because speed serves productivity only when it can be bought within desirable economic and environmental guidelines, this was the goal and challenge of the supersonic cruise technology effort conducted under the NASA Supersonic Cruise Research (SCR) and Variable Cycle Engine (VCE) programs.

CHAPTER 8

Pursuing the Problems and Potential of Supersonic Cruise Flight: The NASA SCR and VCE Programs

As indicated in the previous chapter, a number of difficult technical and environmental problems are associated with commercial supersonic cruise flight. Most of these problems were not satisfactorily resolved during the U.S. SST program. Although an American SST could have been developed with the supersonic technology available at the time, such an SST would not have measured up to the standards of performance and environmental acceptability that the people and the government of the United States demanded. Such an SST would also have been a marginal investment for its airline customers.

Contrary to some views, the cancellation of the U.S. SST program did not establish the insolubility of the problems of commercial supersonic cruise flight or alter the fact that increased speed is an important element of improved productivity, but is a desirable characteristic to be sought and achieved. Consequently, the demise of the SST did not lead to the discontinuation of supersonic cruise technology efforts within the United States. A number of members of the U.S. Congress were concerned about the possible competitive threat of SST efforts in England, France, and the Soviet Union and were in favor of further supersonic technology efforts. The U.S. aeronautical research agency, NASA, viewed the supersonic cruise problems as challenging and difficult, but amenable to solution through a focused research effort.

In early 1972, the Nixon administration directed NASA to formulate a supersonic research program that would provide the technology for a viable commercial supersonic transport. This new program was developed in early 1972 and was originally called the NASA Advanced Supersonic Technology (AST) program. It was to build on the knowledge gained during the U.S. SST program and was to bring a state of supersonic cruise "technology readiness" within 4 years. However, because the opponents of any future American SST program quickly read into the acronym, AST, a program for developing an advanced supersonic transport, the designation was changed to Supersonic Cruise Aircraft Research (SCAR) in 1974. The pace and funding of the program was also cut back so that no "technology readiness" date could be specified. Even with the sharp cutback in anticipated funding, the word "Aircraft" in the title SCAR still gave rise to the spectre of NASA developing an SST. Consequently, the title of the program was changed to Supersonic Cruise Research (SCR) in 1979. This latter designation (SCR) will be used in the balance of this document.

OBJECTIVES AND RATIONALE OF THE NASA SCR PROGRAM

The official objectives of the NASA SCR program were as follows:

> The Supersonic Cruise Research Program has been undertaken to provide a sound data base to support rational decisions in consideration of future civil and military supersonic cruise aircraft.

> The objectives are to define the potential benefits and trade-offs of advancements in aerodynamic efficiency, structures and materials, propulsion systems, and stability and control methods applied to promising advanced supersonic aircraft concepts that also meet environmental requirements. Integration of the technical disciplines will be undertaken, needed analytical tools developed, and wind tunnel and laboratory investigations will be conducted in a closely coordinated effort to provide an advanced technology base.

As this statement shows, the principal objective of the NASA SCR program was to conduct or support disciplinary research on the problems of supersonic flight and to provide advancements in the state of technology. However, in line with one of the lessons learned in the prior SST effort (see Chapter 6), the disciplinary research was to be conducted and assessed, when possible, on the basis of its impact on an integrated supersonic airplane system. This would preclude the expenditure of large amounts on research solutions that had no real application to a practical airplane.

The basic approach of the SCR program, then, was to search for the solution to supersonic problems through disciplinary research. Most of these problems were well known (see Chapter 7), but no satisfactory solution had been found. When the new SCR research suggested a potential solution to a supersonic problem, the applicability of the suggested solution was assessed by determining if it could be integrated into a practical commercial supersonic airplane and mission. If the potential solution withstood the test of the integration exercise, attempts would be made to further validate the solution with wind tunnel tests or hardware construction and tests. If the potential solution could not be integrated, it was discarded, and the disciplinary research teams sought another solution to the problem.

The integrated technology approach of the NASA SCR program was one of its most important aspects. This approach helped to point out the most fertile areas for research and permitted progress to be measured in a quantitative manner.

An important decision in the formulation of the SCR was to conduct the program in a focused manner. SCR was a "line item" in the NASA budget, and this factor promoted an interest and impetus within the aerospace community that a NASA in-house generic research program probably would not have provided. It was perhaps this display of interest by the U.S. government and NASA that prompted some aerospace companies to augment the SCR program with company funding and manpower.

The SCR program was set up to involve all of the NASA aeronautical centers. As the program progressed, it also involved many of the aerospace companies, research organizations, and universities. It was believed that solutions to the supersonic problems could be

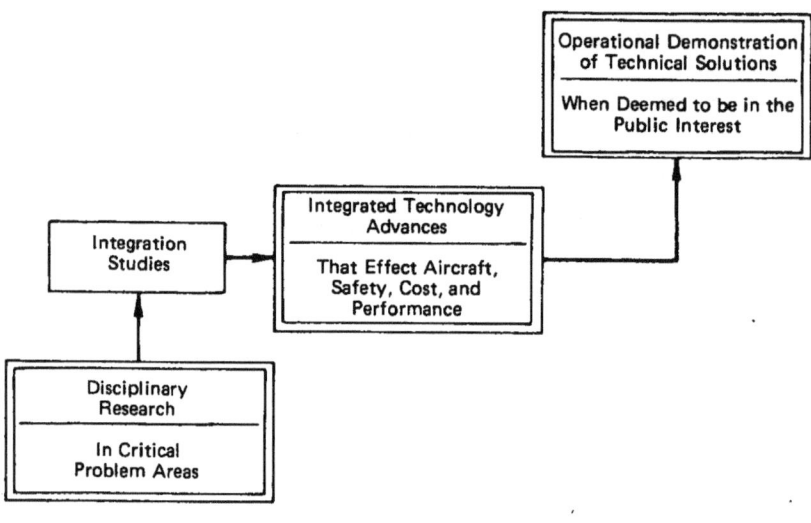

SCR operating plan.

found if there was wide enough participation by those who had experience in the field. Before the SCR program came to an end in 1981, more than 100 separate organizations had been involved in some aspect of the effort.[131]

ORGANIZATION AND ELEMENTS OF THE NASA SCR PROGRAM

The NASA SCR program was conducted under the overall direction of the Office of Aeronautics and Space Technology at NASA Headquarters. Day-to-day operation of the program was the responsibility of the SCR Program Office established at the NASA Langley Research Center. Although the program office had a leader for each major disciplinary research area, the effort required for solving the supersonic cruise problems was in the hands of the research organizations within NASA. The important technology integration function was under the leadership of the SCR Program Office.

NACA SCR AND VCE PROGRAMS

At the outset, the major elements of the SCR program were as mentioned earlier in the objectives statement. The disciplinary research elements were *Aerodynamic Performance, Propulsion, Structures and Materials, Stability and Control,* and *Stratospheric Emissions Impact. Mission Performance Integration,* or *Systems Integration Studies,* was the other major element written into the original SCR Program Plan. Brief descriptions of these elements follow.

AERODYNAMIC PERFORMANCE

This research element was responsible for developing and testing advanced aerodynamic concepts that could be applied to the commercial supersonic transport mission. The element also was responsible for developing and validating advanced analytical techniques for use in

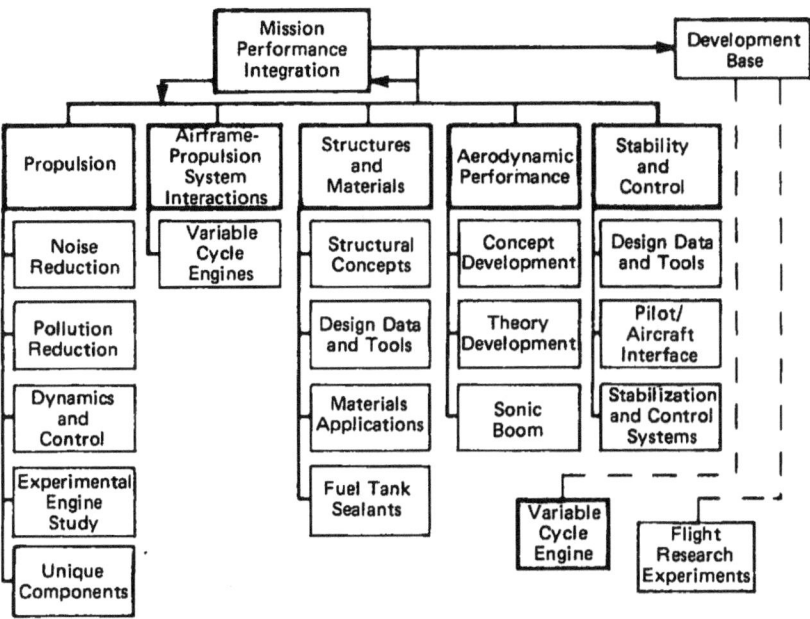

Elements of NASA SCR program.

aerodynamic design and analysis, and for conducting studies to improve the understanding of sonic boom phenomena. Subelements of this research area were Concept Development, Theory Development, and Sonic Boom. Research under the discipline was conducted or monitored mainly by personnel of the Aerodynamic Directorates at the NASA Ames and Langley Research Centers.

PROPULSION

The major goal of this SCR research element was to develop a propulsion system that would efficiently meet the conflicting requirements of subsonic and supersonic operation. Solutions were also needed for the noise and pollution problems that became critical issues in the latter stages of the U.S. SST program. Subelements of the Propulsion program were Engine Studies, Noise Reduction, Pollution Reduction, Dynamics and Control, and Unique Components. This research was conducted or monitored by personnel at the NASA Lewis Research Center.

STRUCTURES AND MATERIALS

Major goals of the Structures and Materials research element were to develop structural concepts and materials that would efficiently withstand the constantly variable load and temperature environment that is experienced during a supersonic flight. The developed concepts and materials would also have to be relatively low-cost if a commercial supersonic transport was to be economically viable. To meet these goals, improved analytical methods for structural design and analysis would be required, as well as improved nonmetallic materials such as fuel-tank sealants and windshields. Subelements of the Structures and Materials program were Structural Concepts, Design Data and Tools, Material Applications, and Fuel-Tank Sealants. Research in this element was under the direction of personnel from the Structures Directorate of the Langley Research Center with participation by engineers and scientists at the Dryden Flight Research and Ames Research Centers.

NACA SCR AND VCE PROGRAMS

STABILITY AND CONTROL

The research responsibility of this SCR element was to develop methods for accurately determining the stability characteristics and control requirements for large, flexible supersonic cruise aircraft and to determine the requirements and problems associated with stability augmentation systems and active control systems. Subelements of this research were Design Data and Tools, Pilot/Aircraft Interface, and Stabilization and Control Systems. Research was directed mainly by the NASA Ames and Dryden Flight Research Centers, with some participation by the Langley Research Center in the Active Control area.

STRATOSPHERIC EMISSIONS IMPACT

The critical research goal of this SCR element was to answer questions that arose during the SST program concerning the pollution of the upper atmosphere by high-flying aircraft. Major questions involved the jet-wake chemistry and how the jet wake propagated and dissipated. There were also questions about the level of stratospheric pollution due to natural causes alone. This research area was critical to the future of supersonic cruise aircraft, but was not a problem unique to such aircraft. The NASA Office of Space Science had been the principal NASA group involved in the problem of stratospheric pollution, and after October 1, 1976, NASA research in this area was funded and managed by that group. All of NASA's aeronautical centers participated in some facet of the investigation of upper atmospheric pollution.

MISSION PERFORMANCE INTEGRATION

The important function of this element of the SCR program was to assess the impact of disciplinary technology advances on the integrated performance of various supersonic cruise aircraft concepts. With this element, it was possible to measure the progress of the technology effect and determine if the technology for an economically viable, environmentally acceptable commercial supersonic transport had been identified. The SCR program was fortunate to get talented, experienced systems integration teams from the Boeing Commercial Airplane

Company, the Lockheed-California Company, and the Douglas Aircraft Company of McDonnell Douglas. An excellent in-house team was also assembled through a NASA nonpersonal services contract with Ling-Temco-Vought. The SCR program was also fortunate to secure experienced propulsion system design groups from the General Electric Company and Pratt and Whitney Aircraft Company.

It might appear that the SCR program was somewhat frivolous for bearing the expense of four systems integration teams and two propulsion design teams. Actually, however, this was a major strength of the program. First, the three major U.S. airplane companies with capability to develop a supersonic cruise aircraft were in the program. (The fourth company, Rockwell International, was brought in later with a propulsion-integration contract.) Second, the two propulsion companies with the capability to build an American supersonic engine were in the program. Third, all of the integration teams had different ideas as to what a commercial supersonic cruise aircraft should look like, what altitude and speed it should fly at, what it should be made of, and how many passengers it should carry. This variation of reference airplane permitted the disciplinary technology to be assessed over a wide range of applications and conditions. Finally, all of the industry teams contributed important disciplinary technology advances during the course of the SCR program in addition to performing their integration functions.

In addition to the foregoing technical factors, the inclusion of the major aerospace companies added an important competitive tone to the program. As a result of this competitive atmosphere, company managements assigned some of their best engineers to the program and supported in-house supersonic cruise research with company funding.

ALTERATIONS TO SCR PROGRAM STRUCTURE

During the course of the SCR program, some additions, subtractions, and alterations were made to the program structure. Although the Stratospheric Emissions Impact research that had been supported by SCR funds was returned to the NASA Office of Space Science in

NACA SCR AND VCE PROGRAMS

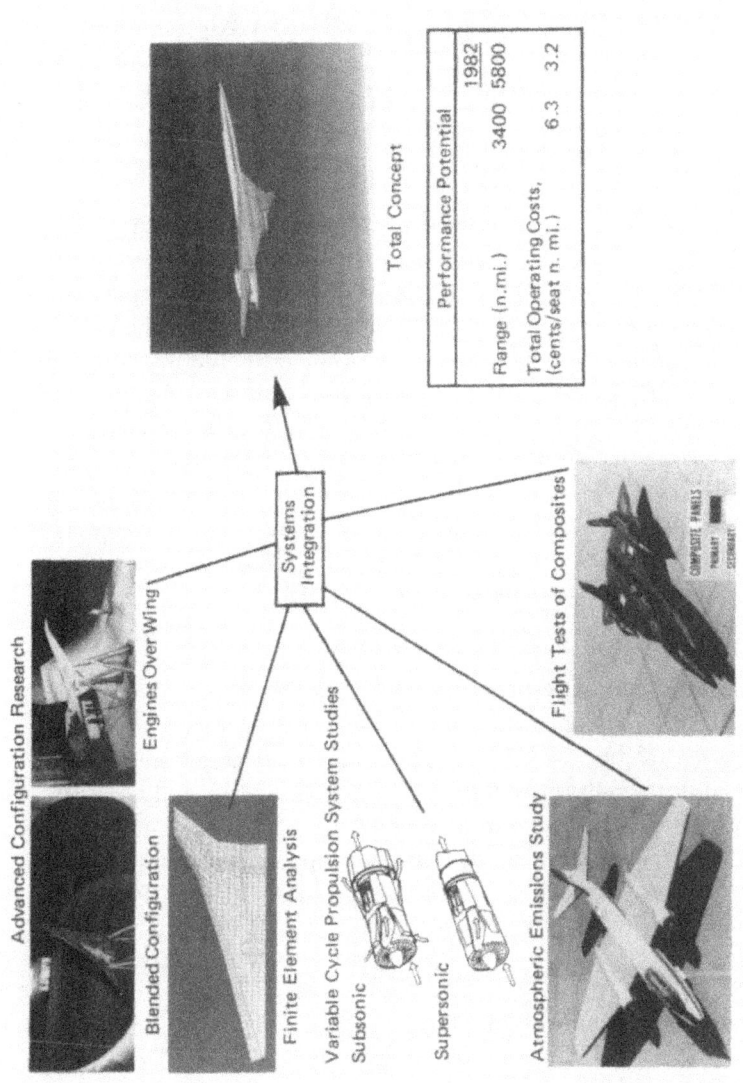

Pictorial representation of NASA SCR program.

1976, the SCR program kept abreast of efforts in the field. Added to the SCR program were the important elements of Flight Experiments and Airframe/Propulsion System Integration. Broken away from SCR, but still closely associated with the effort, was the important VCE program.

FLIGHT EXPERIMENTS

The U.S. Air Force provided NASA with a high-altitude supersonic YF-12 airplane prior to the beginning of the SCR program. Although the airplane did not meet the requirements of a commercial supersonic airplane, it had been useful in supersonic experiments conducted by the Dryden Flight Research Center. It became a useful adjunct to the SCR program, being used to flight-test advanced structural panels, to validate methods for aeroelastic analysis, and to investigate the problems associated with airframe/propulsion system interactions at supersonic speeds. The YF-12 was also quite useful in the study of control system problems and in the development of control system concepts. Personnel at the NASA Dryden Flight Research Center supervised the flight tests and experiments involving the YF-12.

AIRFRAME/PROPULSION SYSTEM INTEGRATION

In 1977, after the SCR program had identified a technology base that could lead to a viable SST, it became apparent that a more detailed look should be taken at the important interfaces between the airplane and its propulsion system (i.e., the engine inlets and nozzles). Accordingly, the Lewis Research Center, the SCR Program Office, and the mission integration contractors developed a research program for studying these interfaces. Personnel from the Lewis Research Center monitored this effort, which was initiated in 1977.

VARIABLE CYCLE ENGINE PROGRAM

During the first 3 years of the SCR program, the propulsion system contractors, General Electric and Pratt and Whitney, identified engines that showed promise of efficiently meeting the subsonic/supersonic requirements of an SST mission. The two engines accomplished the conflicting requirements of subsonic/supersonic flight with variable

USAF Lockheed YF-12 served very useful function with flight tests for SCR and VCE programs.

features that permitted the engine to change operating characteristics during flight. To focus effort on these two promising "variable cycle" engines, a separate Variable Cycle Engine program was established in 1976. Although the VCE program became a separate "line item" in the NASA budget, it was closely associated with the SCR program. A program office at the NASA Lewis Research Center was formed to manage this effort.

DISSEMINATION OF SCR TECHNICAL INFORMATION

In many respects, SCR technical information was disseminated in the same manner as that of other NASA programs. Nearly 1000 reports and presentations resulted either directly or indirectly from research supported by the SCR program.[132, 133] In addition, formal annual reviews of the disciplinary research and mission integration results and two major NASA conferences* were held in 1976 and 1979. These conferences were well attended by members of the aerospace, military, and academic communities.

Certain elements of NASA SCR technology and data were controlled through the use of the heading "For U.S. Government Agencies and their Contractors Only." A unique feature of the SCR program, however, was the rapid and almost simultaneous dissemination of this technology and data to most of the major military and civilian airplane and propulsion manufacturers. This rapid dissemination was possible because four major airplane and two major propulsion manufacturers were involved in the SCR program. Boeing, Lockheed, and Douglas were prime mission integration contractors, and Rockwell was involved in a propulsion integration contract. General Electric and Pratt and Whitney provided propulsion data to all of the airplane contractors. Each of these contractors presented their disciplinary research and mission integration results at the annual SCR reviews. Except for the first few years, when neither propulsion contractor wanted the other

---

*The SCAR Conference (NASA CP-001, 1976) and Supersonic Cruise Research '79 (NASA CP-2108, 1980).

present at his review, the annual meetings were open to representatives from all companies. During the course of the SCR program, Boeing, Lockheed, and Douglas conducted a considerable amount of SCR-related technology effort with their "in-house" funding and manpower. All of the companies reported the results of these proprietary data at meetings attended by the other contractors.

An unusual spirit of cooperation existed among the contractor teams even though each was competing for a larger share of the SCR funding. For example, Rockwell, which had bid unsuccessfully to become an SCR mission integration contractor, provided a meeting place for the three winning teams to review their progress. On another occasion, Douglas provided the SCR office with some valuable proprietary noise suppressor data for use in a program for supporting the Federal Aviation Agency in setting noise rules for supersonic aircraft.

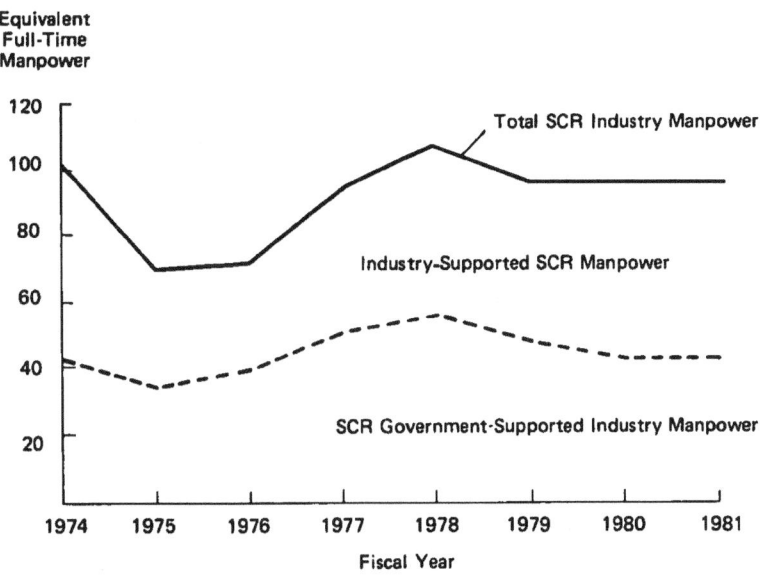

Major airplane contractors supported SCR and VCE with "in-house" funding and manpower.

113

SUPERSONIC CRUISE TECHNOLOGY

SCOPE OF NASA SCR PROGRAM

Some opponents of the SST saw the SCR program as a surreptitious attempt to resurrect the SST that had just been canceled by the U.S. Congress. This feeling became even more prevalent when the program brought the major airframe and propulsion manufacturers into the systems integration function. Further misunderstanding resulted from the use of reference supersonic transport airplanes and missions for measuring the progress of the technology.

These concerns, though probably natural, were completely unfounded. The SCR program was no more than an effort by NASA to meet its chartered commitment to provide technology that would ensure American supremacy in civilian and military aircraft. The $300 million NASA supersonic technology effort proposed in response to the Nixon administration directive was by no means an SST development program. Rather, it was a minimum 4-year proposed program for providing the technology base for possible future use in the development of an SST. The decision to involve all of the major airframe and propulsion companies was made essentially to get as much help and as many ideas as possible for solving the very difficult supersonic problems. The use of practical supersonic cruise aircraft for assessing technology was dictated by experience, which had shown that an advancement in one supersonic discipline means little unless it can be integrated with the other supersonic requirements.

The cutback of anticipated SCR funding in the second year and the eventual support level of around $10 million per year made it impossible to set or shoot for any "technology readiness" date. However, this cutback in funding support did not alter the supersonic cruise aircraft problems or change the SCR objective of finding solutions to these problems. The scope of the program had always been to establish a technology base that could be used to develop a viable SST whether it was a 4-year program or an indefinite one. The next two chapters describe some of the efforts of the SCR program to establish this technology base.

CHAPTER 9

Progress in Supersonic Cruise Technology Since 1972

In the years since 1972, the U.S. Congress has supported a number of research programs on supersonic cruise technology. In addition to the NASA SCR and VCE programs, there were the Department of Transportation (DOT) SST Follow-On Program, the Climatic Impact Assessment Program (CIAP), the High Altitude Pollution Program (HAPP), and the NASA Emissions Reduction Research and Technology Program. The DOT SST Follow-On Program supported the continuation or completion of research on several technology items associated with the then-canceled SST. Some of this research found important application to later military and commercial airplanes, and some of the technology elements became the basis for other NASA and industry programs.[134] Results from the CIAP and HAPP efforts provided a substantially improved understanding of atmospheric pollution, and the NASA Emissions Reduction Program provided means for reducing the pollution from jet engines.[135]

The DOT SST Follow-On, CIAP, HAPP, and NASA Emissions Reduction efforts were directed either to the solution of a specific supersonic cruise problem or to the accumulation of data that would shed light on the magnitude of the problem. All of the programs contributed greatly to the supersonic technology, but none had the overall responsibility for developing the base for a viable supersonic cruise aircraft. This responsibility was given to the NASA SCR program when it was established in 1972. The SCR effort was to assimilate the results of the other supersonic programs with whatever new research was necessary for developing the desired supersonic technology. The VCE offshoot program bore the same responsibility in the propulsion area.

Supersonic technology efforts in the United States after 1972 involved several hundred individual, but coordinated, research efforts. The SCR and VCE programs alone had more than 100 research areas, many of which involved subareas and phases. Some of these research areas were not expected to provide far-reaching technology advances, but were needed to furnish data banks that could be used to make a dramatic breakthrough. In other research areas, such as engine noise, any advance would be dramatic because of the "front line" importance of the research area. Although this chapter and the next deal mainly with progress in these "front line" issues, they also mention the important "data gathering" efforts, when appropriate. Liberal excerpts are taken from Driver's excellent paper entitled *Progress in Supersonic Cruise Technology*.[136] This chapter examines progress in environmental issues, aerodynamics, structures and materials, propulsion, and configuration concepts/integration.

PROGRESS IN ENVIRONMENTAL ISSUES

Environmental issues such as noise, sonic boom, and pollution had a lot to do with the cancellation of the U.S. SST program and with the failure of supersonic cruise aircraft to assume a prominent role in air transportation. These issues could also be responsible for a reticence to consider the SST as a future transportation system. Certainly, any future SST program would have to answer the environmental questions before any action could be taken. Although progress has been made on many of the environmental issues, some questions remain.

UPPER ATMOSPHERIC POLLUTION

As indicated in Chapter 7, the SST brought fears of ozone depletion in the upper atmosphere and of an increase in the incidence of skin cancer on the Earth. The CIAP study showed these concerns to be overstated, and the later HAPP program indicated a possible increase in ozone rather than a depletion.[137] Nevertheless, there is still a lack of complete understanding of the chemistry of the upper atmosphere and the effect of jet-engine emissions on it. Consequently, programs for reducing emissions are still needed.

Among the programs that attempted to provide the technology for reduced engine emissions were the NASA Emissions Reduction Research and Technology program and elements of the SCR Propulsion and VCE programs. These programs identified concepts that could reduce nitrous oxide (NO_x) emissions to less than one-third the levels of first-generation SSTs and provide for substantial reductions in carbon monoxide (CO) and unburned hydrocarbons (THC).[138] Although further study of the chemistry of the upper atmosphere could indicate that these concepts (complicated advanced burners) might be unnecessary, their validation for subsonic or supersonic use is desirable.

SONIC BOOM

As indicated in Chapter 7, the sonic boom is a direct result of supersonic flight, and no means is available for eliminating this accompaniment of such speeds. As mentioned in Chapter 4, NASA has conducted and supported a massive program for improving the understanding of sonic boom and for reducing its impact by design. None of these programs have given a reliable hint, however, that this disturbance can be completely abolished. Consequently, the major goal of the SCR program was to establish a level of public acceptability of the sonic boom to guide research on design methods for reducing the disturbance. Plans for this program were put in limbo when the anticipated funding for the SCR effort was reduced from more than $40 million in the second year to approximately $10 million. Because a rule was in effect that prohibited supersonic overland flights of commercial supersonic aircraft, it was difficult to justify a large sonic boom effort in light of the cut in the overall program.

It is quite possible that a level of sonic boom can be found that is acceptable to most of the public and that such a level could be achieved through the design methods developed by NASA and others. From the

117

standpoint of the SCR program, it was decided that the most feasible way to accomplish this goal was to develop the technology for an outstanding overwater SST, provide the technology that would permit the flexibility of efficient subsonic overland flight, and then use this vehicle to explore the sonic boom acceptability question. Although this did not represent a direct frontal attack on sonic boom, it appeared to make best use of the funding at hand.

ENGINE NOISE

Engine noise, a necessary environmental issue with all modes of engine-powered airplane flight, was a particular problem with the SST. Although many of the subsonic jets were as noisy, they appeared to be much more necessary than supersonic transports. Much progress was being made on subsonic jet transport noise, and nothing less could be tolerated for an SST. Consequently, engine noise was a primary research element of the NASA SCR program and, later, the NASA VCE program.

During the course of the SCR and VCE programs, some exciting technologies were discovered that showed promise for dramatically reducing the projected noise of supersonic cruise aircraft. The first of these technology advances was the inherent coannular noise reduction of the variable cycle engines under consideration in the SCR propulsion program at the NASA Lewis Research Center. These engines utilized an "inverted jet" velocity profile (i.e., a high-velocity outer jet stream exhausted through a high-radius-ratio annulus and a lower velocity jet stream exhausted through an inner nozzle). This system was the opposite of the normal turbofan engine, which has an inner high-velocity stream and an outer low-velocity stream. The inverted velocity profile (IVP) coannular nozzles were extensively tested and validated in a series of model-scale tests, tests on a complete running engine, and free-jet tests to check the effects of forward velocity.[139] These tests verify a significant noise reduction (approximately 7 EPNdB) relative to a fully mixed conical nozzle at the same specific thrust and mixed-pressure ratio.[140]

118

Coannular noise tests in Boeing anechoic facility (courtesy of Boeing).

The second important engine noise development in the SCR program was the mechanical suppressor technology validation by the Douglas Aircraft Company. This effort, perhaps the most complete one under the partial aegis of the SCR program, involved: (1) small-scale suppressor development tests by Douglas at a Douglas facility; (2) spin tests of the suppressor on Roll Royce's rotating-arm rig in England; (3) flight tests of the mechanical suppressor nozzles on a Rolls Royce Viper 601 engine installed on a Hawker Siddeley HS-125 aircraft modified to accept the engine/nozzle/suppressor/treated-ejector combination; and finally (4) tests of the engine, nacelle, and nozzles on

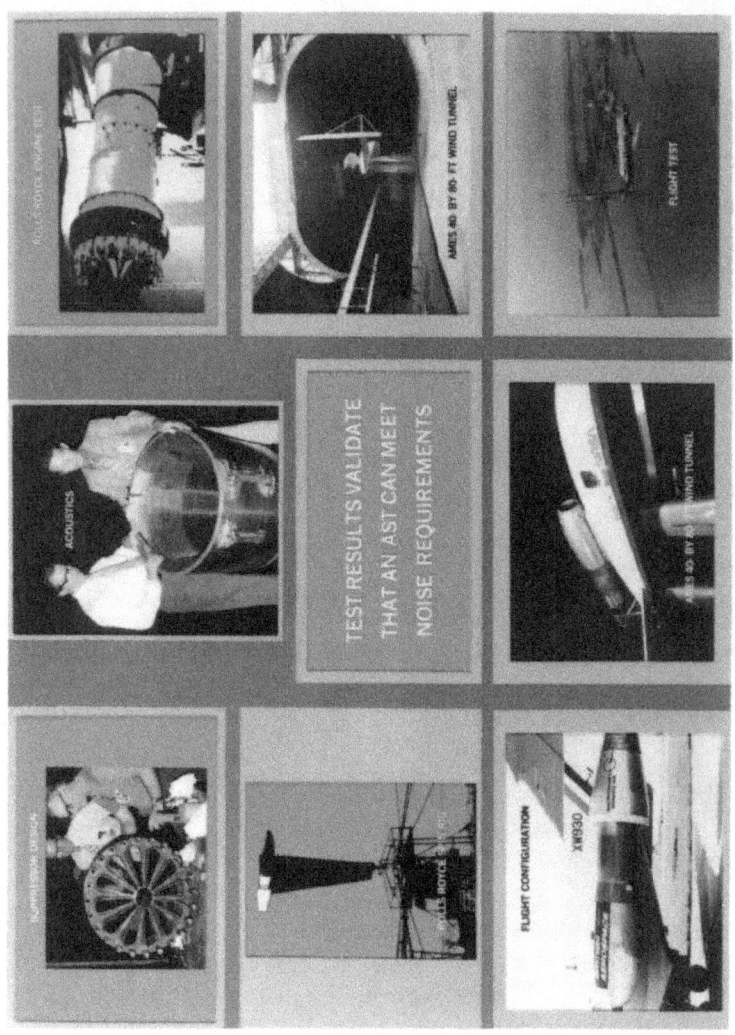

In-depth analytic, experimental, and flight test program proves McDonnell-Douglas advanced noise suppressor.

a simulated fuselage in the NASA Ames Research Center 40- by 80-foot wind tunnel. The combination of tests indicated that the Douglas multitube/lobe retractable suppressor with acoustically treated translating ejector would provide 12 to 16 EPNdB noise reduction relative to a conic nozzle at equal thrust and a thrust loss of 4.5 percent relative to a conic nozzle at takeoff.[141, 142] This suppressor performance is much better than that of previous suppressors, and the results provide real hope that the mechanical suppressor can be a powerful noise-reduction device on future supersonic airplanes. It remains to be seen whether suppressors of this type can be integrated on the engine and if they can withstand the operating pressures and temperatures at takeoff conditions.

A third important noise technology development of the SCR program was the concept of acoustic shielding. Two methods of noise shielding were considered. First, the Lockheed Company placed the engines in an over/under arrangement on the wing and used the wing and fuselage to shield some of the noise from the ground. This approach promised a noise reduction of 3 to 5 EPNdB compared to that of placing all engines under the wing in the conventional manner.[143] The other approach to noise shielding made use of a thermal acoustic shield (i.e., a high-temperature low-velocity gas stream, partially surrounding a high-velocity central jet exhaust). This concept was borne out in theoretical predictions and experimental tests conducted by NASA, the Boeing Company, and the General Electric Company. Although this noise reduction concept is still being investigated, the effect is believed to be associated with the reflection and refraction properties of the high-temperature acoustic shield.[144]

The use of "minimum noise" flight profiles during terminal area operations was the fourth noise reduction concept to show promise in the SCR program. This approach essentially makes use of advanced operating procedures such as thrust modulation during ground roll and takeoff, engine cutback at optimum "noise" altitude, thrust cutback to optimum "noise" level, and configuration changes during takeoff other than raising the landing gear. Making use of these procedures during takeoff and using decelerated approaches and increased glide slopes

121

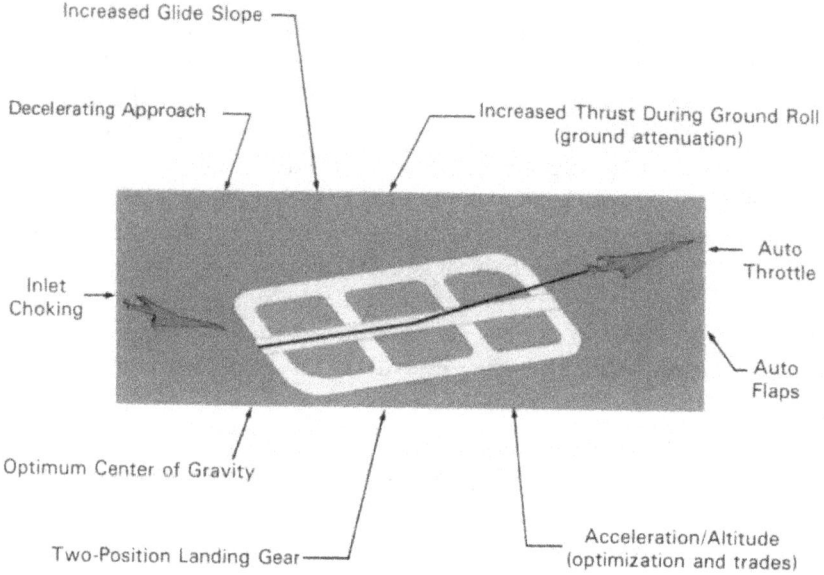

Advanced operating procedures for noise reduction.

during landing operations can lead to substantial decreases in engine noise exposure.[145] Unfortunately, current Federal Air Regulations (FAR) do not permit these advanced procedures. Several elements of the SCR and supporting programs could make it possible to certify these procedures in the future. The YF-12 flight research program included an effort to develop a cooperative airplane/engine control system that would permit the fine degree of control required of the advanced operating procedures.[146] In addition, the SCR stability and control element supported visual piloted simulator studies of the use of these procedures on an advanced supersonic transport. It was determined that the advanced operating procedures did not compromise flight safety.[147]

The substantial noise reduction potential of the methods identified in the NASA SCR and VCE programs give rise to some optimism about the potential noise characteristics of future supersonic cruise aircraft.

122

PROGRESS IN TECHNOLOGY SINCE 1972

Although probably not any of these separate methods will provide the desired noise levels, a combination of methods appears to be possible. The basic underlying principles of most of these noise reduction methods have been validated in some form of experimental program, in sharp contrast to the noise technology situation that existed at the close of the SST program in 1971. At that time, the only readily available noise reduction technology was a vastly oversized propulsion system that had an extremely deleterious effect on the entire airplane concept.

PROGRESS IN AERODYNAMICS

As indicated in Chapter 4, the status of the aerodynamic technology for supersonic cruise vehicles had been brought to a high level by the close of the SST program in 1971. Not only had accurate supersonic design and analysis methods been developed, but an advanced arrow-wing configuration, the NASA SCAT-15F, had been introduced. The supersonic cruise aerodynamic efficiency (lift/drag) of the SCAT-15F was 25 percent greater than the previous state of the art. In spite of this advanced level of aerodynamic technology, however, the supersonic and supporting aerodynamic activity was near its peak during the SCR program. It was found that improvements to the design and analysis methods could still be made,[148] and it was also indicated that entirely new methods might be needed to predict detailed aerodynamic parameters, such as pressure distributions, at critical structural and control design conditions.[149] Meanwhile, the highly efficient supersonic arrow-wing concept stood unused because of questionable subsonic aerodynamic characteristics, which also gave rise to relatively poor noise characteristics.

Although some effort was spent on improving supersonic aerodynamic efficiency during the NASA SCR program,[150, 151] the principal thrust of the aerodynamic program was to find potential solutions to the low-speed aerodynamic problems of the highly swept arrow-wing concept. This thrust was doubly important because Lockheed and Douglas had reference configurations that employed this

SCR program also considered improvements in high-speed aerodynamics. SCR/ McDonnell-Douglas model tested at NASA Ames and Langley Unitary wind tunnels.

concept, and at one point, Boeing proposed an arrow-wing supersonic technology demonstrator airplane.

For investigating how to improve the low-speed performance of highly swept arrow-wing concepts, the SCR program supported a massive model construction program that included large models of a blended concept, the NASA reference SCAT–15F derivative, the McDonnell-Douglas Mach 2.2 concept, and a number of generic arrow-wing models. The experimental programs for testing this myriad of models were conducted mainly by personnel in the NASA Langley Research Center 30- by 60-foot tunnel, the 7- by 10-foot tunnels, and the VSTOL tunnel. Many technicians and engineers were involved, and nearly every conceivable test variable was considered. As a result of this concerted effort, the low-speed aerodynamic efficiencies of arrow-wing concepts were improved,[152] solutions for some of the stability and control problems were found,[153] and a substantial arrow-wing low-speed data base was established.[154, 155]

Computer application to rapidly determine:

- Supersonic wave drag
- Drag due to lift
- Interference

Has led to:

- Highly swept, arrow-wing planform (NASA SCAT 15-F)
- Highest values of L/D measured experimentally (L/D = 9 @ M = 2.7)
- 20 to 30 percent longer ranges for supersonic transport configurations
- Widely used throughout the aircraft industry for both civil and military application

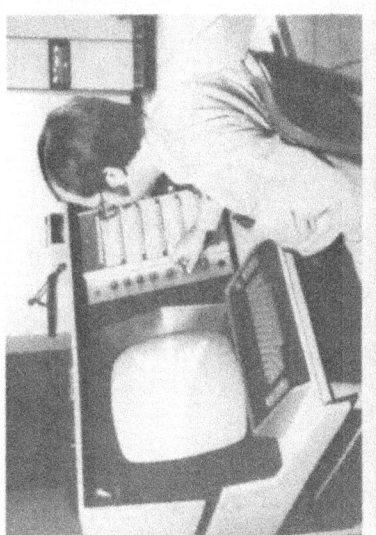
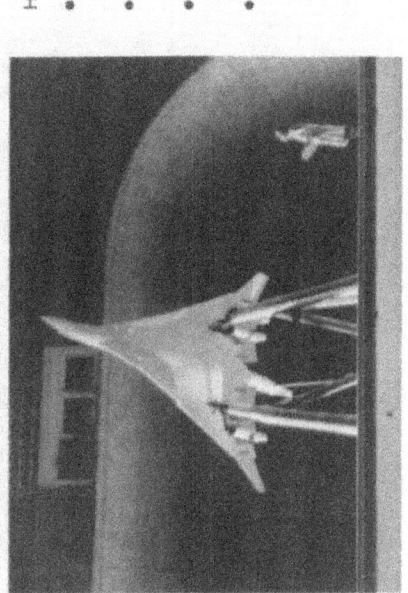

Aerodynamic design methodology improvements are also an SCR goal.

SCR aerodynamic test program — low-speed tests of McDonnell-Douglas arrow-wing concept in Langley Research Center 30- by 60-foot full-scale wind tunnel.

Not only did the low-speed performance of arrow wings improve during the SCR program, but the low-speed and high-speed performance of the Boeing–300 delta-wing concept improved 16 percent as well.[156] The combined improvements in aerodynamic analysis methods and aerodynamic efficiencies, accomplished during the SCR program, essentially brought this research discipline into a state of "technology readiness" for the development of viable supersonic cruise aircraft.

PROGRESS IN STRUCTURES AND MATERIALS

As indicated in Chapter 7, the structures and materials technology for building a supersonic cruise aircraft was available in the 1950s. The technology for building an acceptable supersonic transport was not available, however, and much of the technological shortfall was in the

Tethered free-flight tests of SCR arrow-wing model in Langley Research Center full-scale wind tunnel. Findings: (a) model longitudinally stable over test α range; (b) coupled roll and yaw control provided effective lateral control up to $\alpha = 27°$.

Another of SCR's test efforts to improve low-speed aerodynamics of highly swept arrow wings — blended model in Langley full-scale wind tunnel.

SCR variable-sweep model in 30- by 60-foot wind tunnel.

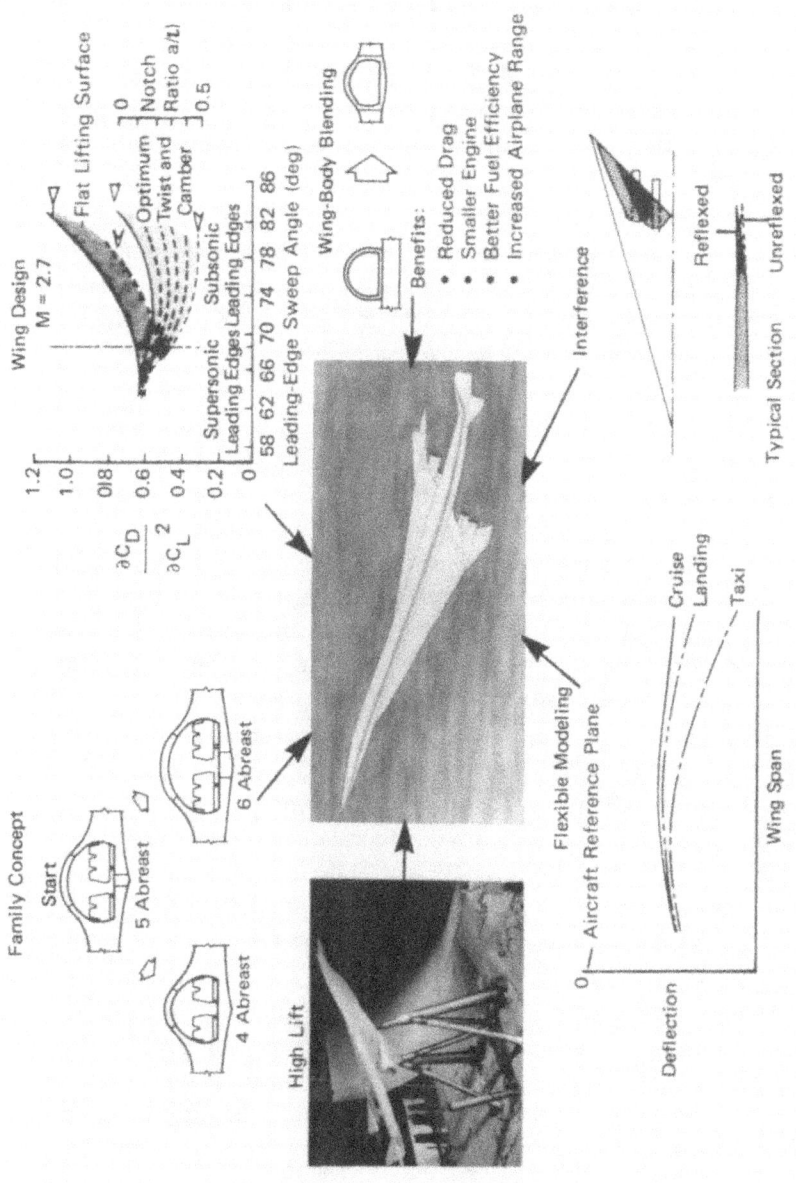

Summary of SCR aerodynamic research effort.

structures and materials discipline. Mentioned particularly were the inability to rapidly and accurately assess the effect of design changes on the complex flexible structure of a supersonic transport and the absence of adequate low-cost structural concepts and fabrication techniques. Another critical need, which was not mentioned, was for a fuel-tank sealant material that would retain its integrity in supersonic flight applications.

The structures and materials effort that evolved out of SCR support concentrated on these problems, but it also provided important data bases on supersonic loads and landing loads. An in-depth summary of this program is presented by Richard Heldenfels of the Langley Research Center in *Supersonic Cruise Research '79.*[157]

The NASA effort to improve methods for structural design and analysis has spanned the area from attempts to improve inputs, such as unsteady aerodynamics,[158] to research for improving the massive combination of programs that provide the design and analysis capability.[159] This effort has been successful as "advances continue to be made in the structural analysis and design area with the result that computational procedures are available now to design quickly a vehicle structure that meets the requirements for strength, divergence, and flutter with active controls included. This can be done accurately and early enough in the design process to avoid costly changes during detail design."[160]

Research efforts to develop low-weight, low-cost structural concepts and low-cost fabrication techniques also covered a wide range of activity, from the time/temperature/stress analysis of composite materials at supersonic flight conditions,[161] to the fabrication of relatively large superplastically formed titanium panels.[162] This latter fabrication process, the superplastic forming and concurrent diffusion bonding (SPF/DB) of titanium, appears to offer much promise for weight and cost reductions on future supersonic aircraft.

The SPF/DB process, which was being investigated by Rockwell International during the B-1 bomber program, essentially involves the heating of a sheet or sheets of titanium in a mold until the titanium reaches a malleable temperature. A gas is then injected into the mold, and the titanium is either blown into a shape prescribed by the mold or

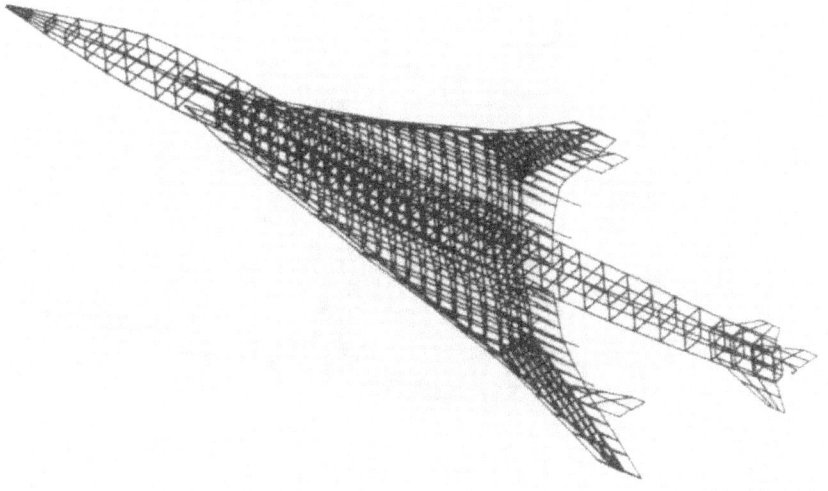

Advanced structural methods permit the use of many structural elements and degrees of freedom for more accurate analysis.

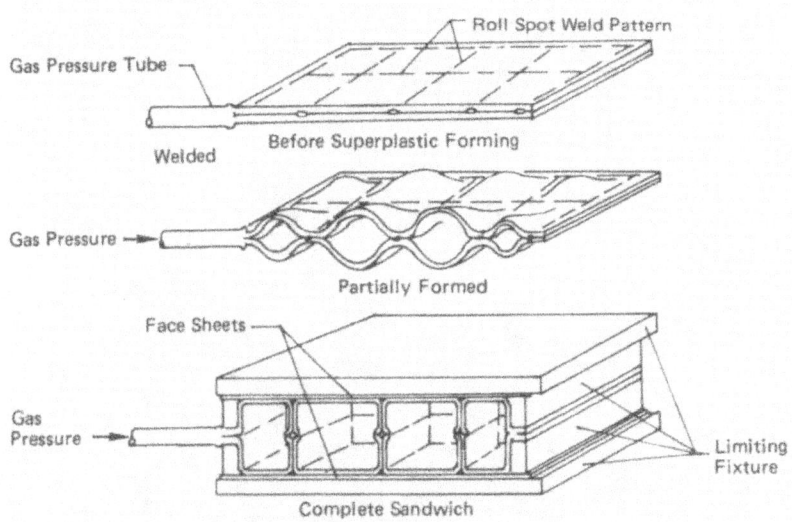

Superplastic forming and concurrent diffusion bonding (SPF/DB) of titanium — an exciting process that permits complex structural elements to be blown into shape.

131

SPF/DB structural panel fabricated by McDonnell-Douglas (courtesy of Douglas Aircraft).

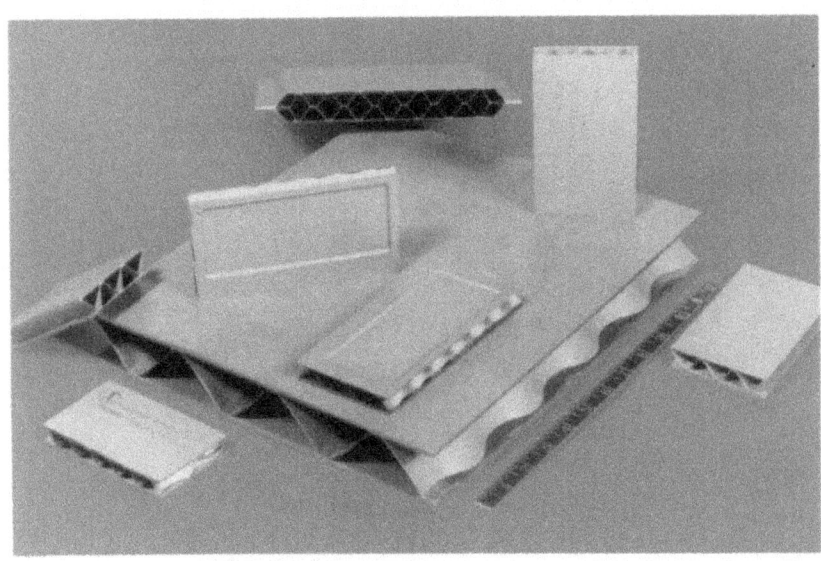

Array of complex structural elements formed by Rockwell International with the SPF/DB process (courtesy of Rockwell International).

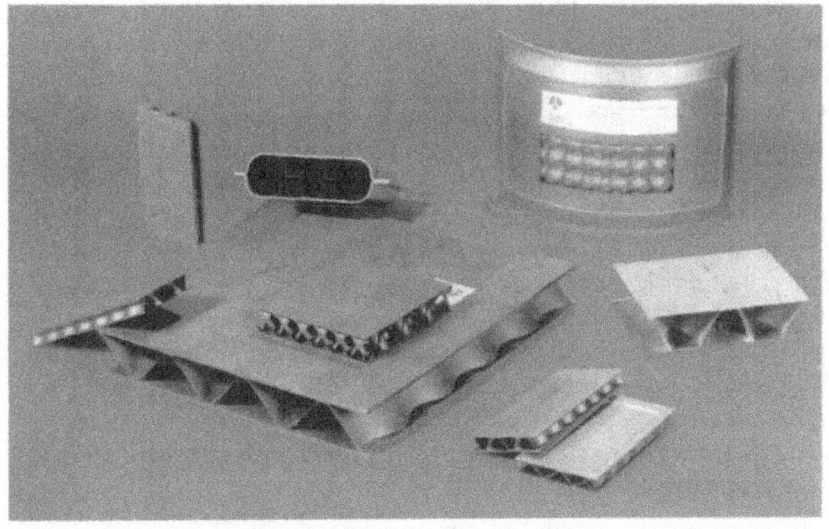

More Rockwell International SPF/DB samples (courtesy of Rockwell International).

bonded to another titanium sheet with a bond of parent-metal strength. There seems to be no limit to the structural elements that can be formed by this process, and no fasteners are required within the structural element. An analysis of the use of this process in the design of a supersonic transport wing showed significant advantages over the aluminum brazed titanium honeycomb that was used in the Boeing U.S. SST design.[163]

The Douglas Aircraft Division of McDonnell-Douglas has made further studies on the SPF/DB process. In this SCR-supported effort, four flat titanium sheets are placed in a mold, heated to plastic-metal temperatures, blown into shape, and diffusion-bonded together. Sandwich cover panels for the wing of a reference airplane configuration made in this manner, combined with similar methods for the wing internal structure, have reduced wing weight by about 7 percent. In addition, a change in the fuselage structure from titanium skin stringer to a titanium SPF/DB sandwich construction reduced the fuselage weight by about 22 percent.[164]

Still another fallout of the SCR program is a technique that makes use of fibers to reinforce the basic titanium structure. This process promises further dramatic gains in structural efficiency and further cost reductions.[165, 166]

The final major problem area in the structures and materials research discipline was the need for a suitable fuel tank sealant for high-speed, high-temperature operation. This was a problem in the SST program and was the subject of study in the DOT SST Follow-On Program. During the SCR program, this research area, directed by the NASA Ames Research Center, produced elastomers based on a polymeric heterocyclic fluoroether that could prove to be a satisfactory fuel tank sealant for supersonic airplanes. This material has shown excellent thermal stability and low-temperature flexibility. It retains the stable characteristics in the presence of jet fuel and resists oxidation at high temperatures.[167]

In summarizing the technology effort in the SCR structures and materials effort through 1979, it has been indicated that "This technology can be used to design safe and durable structures of reduced weight and cost to improve the performance and economics of future supersonic cruise aircraft."[168] This statement was amplified in a Boeing report which stated that the "Results of this work have been very encouraging and, in particular, have made it possible to define configurations of high aerodynamic potential. These configurations have been considered practical only because of the design refinements possible by the successful development of structural technologies. . . . [169]

PROGRESS IN PROPULSION

Engine noise and pollution are serious propulsion problems and were considered as such in the SCR program. If these two issues are considered in the environmental impact area, as has been done in this document, the main remaining and still critical issues confronting the SCR propulsion discipline were the development of propulsion systems that could efficiently meet the contradictory operating requirements of subsonic and supersonic speeds, and then to develop the component

PROGRESS IN TECHNOLOGY SINCE 1972

Concept

- Combination of Advanced Titanium and High Strength Fibers

- Joined by
  – Diffusion bonding
  – Brazing

- Into a Unitized, Load Carrying Structure

Applications

- Stiffness Critical Structure

- Compression Buckling

- High Axial Loads

Benefits
(Ref. Titanium Plate)

- Up to 65% Reduction in Weight
- Up to 24% Savings in Cost

Exciting new structural process makes use of titanium reinforced with fibers.

SCR structures effort includes aeroelastic tests of Boeing SCAT–15F arrow-wing model (courtesy of Boeing).

135

SUPERSONIC CRUISE TECHNOLOGY

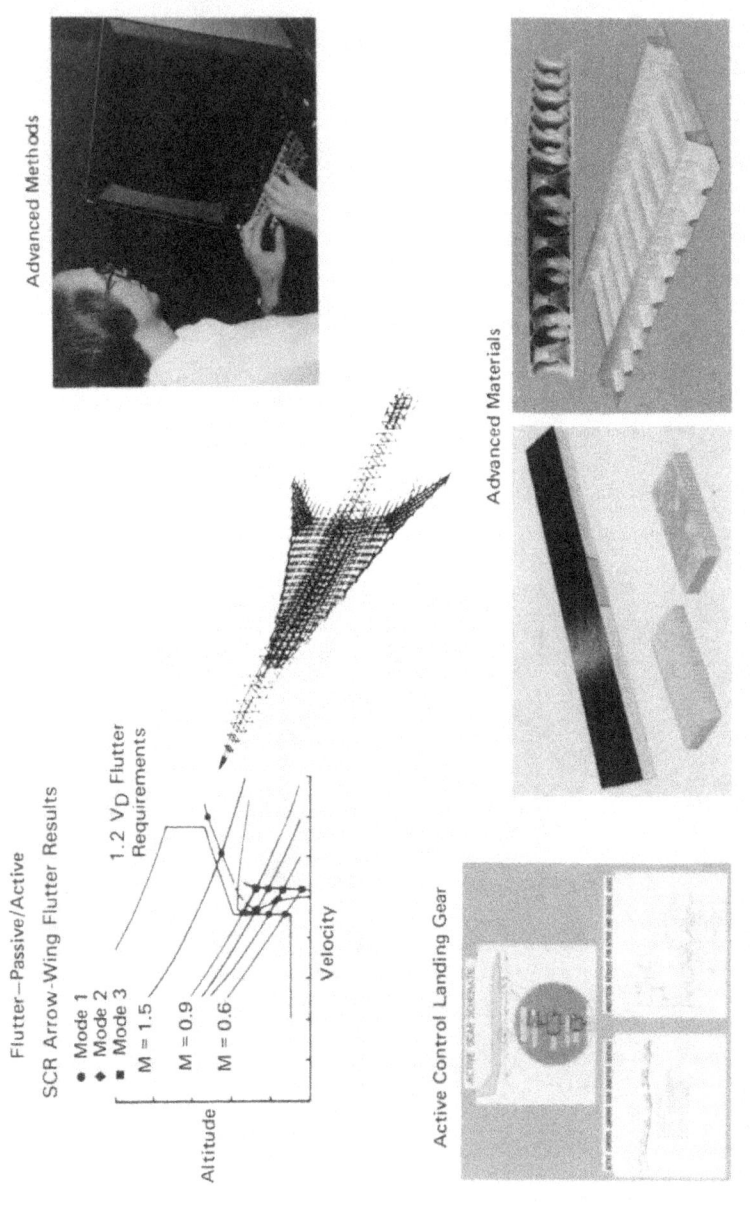

Summary of SCR structures effort.

136

technology that would make these engines possible. Since the noise and pollution progress has already been discussed, only progress in the latter areas will be discussed here.

PROPULSION STUDIES

The SCR approach to the supersonic transport propulsion problem was to take a fresh look at all the propulsion cycles that would have any chance of meeting the demanding supersonic mission. These early studies were conducted by the NASA Lewis Research Center with the assistance of the propulsion system manufacturers, Pratt and Whitney and General Electric,[170] and making use of the SCR mission integration contractors, Boeing, Douglas, and Lockheed.

On the basis of these preliminary SCR studies, it was soon apparent that a successful SST would require an engine cycle that would approach the performance of the turbojet engine at supersonic speeds and the turbofan engine at subsonic speeds. Subsequently, such engines were identified, and the evolution of the Variable Cycle Engine pro-

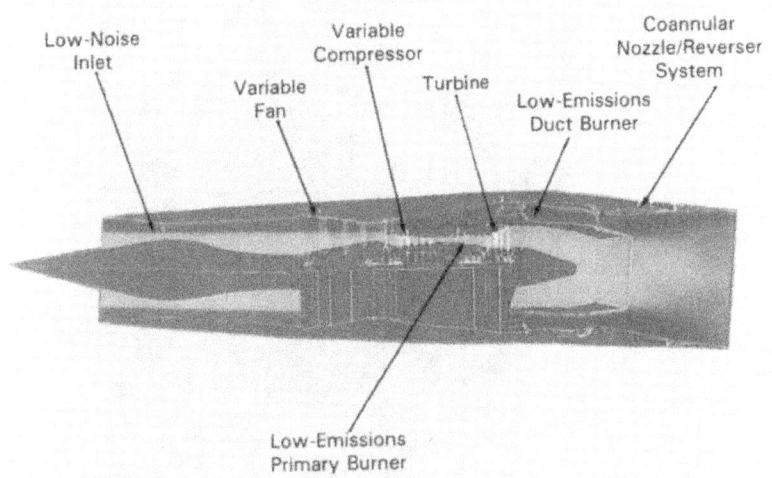

Flow diagram of Pratt and Whitney variable-stream control engine (VSCE) in SCR/VCE program (courtesy of Pratt and Whitney).

General Electric testbed engine for proving the principles and components of G.E.'s VCE (courtesy of General Electric).

PROGRESS IN TECHNOLOGY SINCE 1972

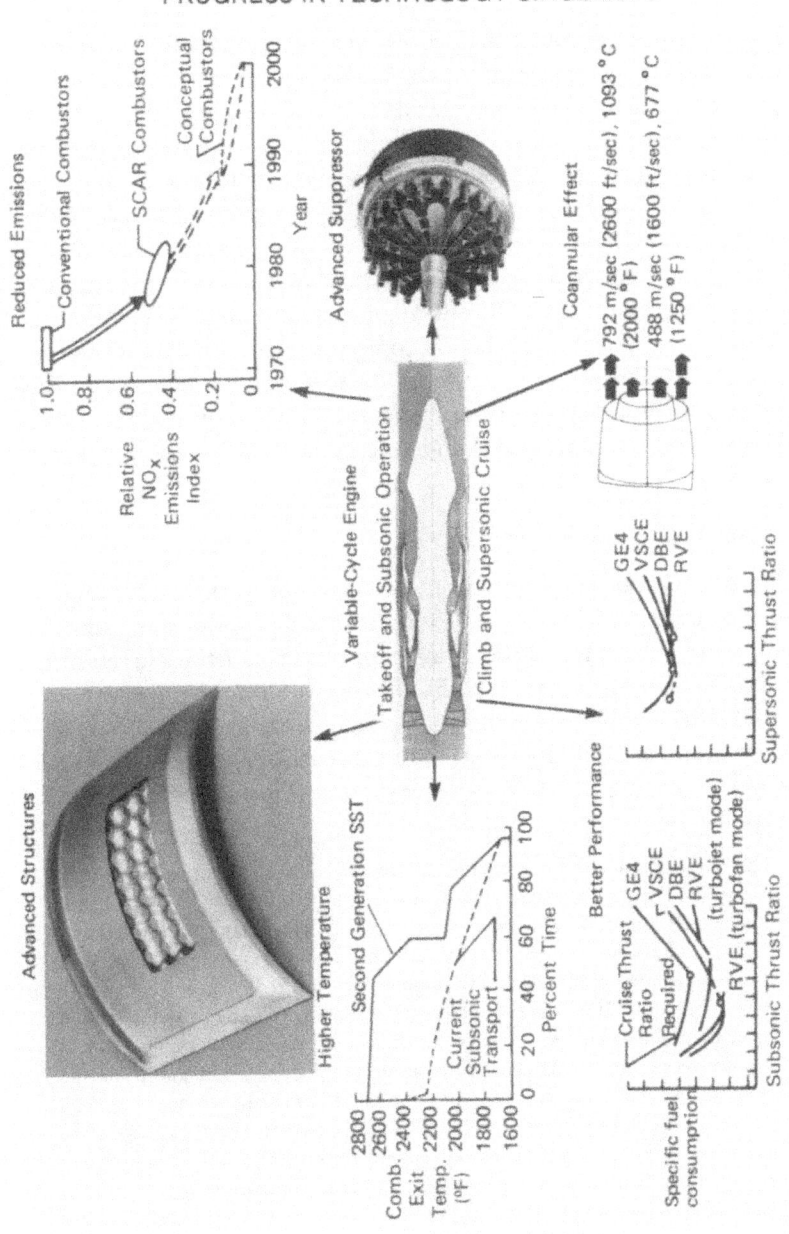

SCR/VCE propulsion summary.

139

gram had begun. Through the process of evaluation by the mission integration teams on a realistic supersonic cruise aircraft mission and on a series of realistic supersonic airplane configurations, the technical shortcomings of the engine cycles could be identified and progressive alterations could be made. Through this refinement process, by 1976, each propulsion contractor had identified a "variable cycle engine" that showed promise. At this stage, the NASA VCE program was formed to develop and conduct programs for validating the critical features of these two engines. This VCE effort continued to be closely tied to the SCR program, and the propulsion companies were under contract not only to the VCE program but also to the SCR mission integration contractors. This situation led to rapid dissemination of technical information between the two programs and among the airframe and propulsion contractors.

VARIABLE CYCLE ENGINE ·

The most important propulsion development in the SCR program was, of course, the "variable cycle engine" concept and the apparent inherent noise advantage of the dual coannular exhaust discussed in the previous section on "Progress in Environmental Issues." The details of the VCEs that evolved from the SCR program and were followed in the VCE program are not important to this discussion, but only the overall function of the engine. A good description of this function is provided by Sigalla:

> The need for variable-cycle engines in relation to the problem of designing a successful SST had been recognized for a long time. But it was only as a result of the SCR program that coordinated research by airplane manufacturers, engine manufacturers, and NASA technologists led to the mechanical and thermodynamic definition of such engines by Pratt & Whitney Aircraft and General Electric. It should be noted that a variable-cycle engine is not defined by any specific mechanical scheme. Rather, it is defined by its ability to meet a set of requirements aimed at eliminating the poor subsonic and transonic performance of supersonic engines designed for higher Mach numbers without affecting adversely the supersonic cruise performance of those engines. Such requirements are high supersonic cruise performance with low specific fuel consumption and high specific thrust (comparable to a dry turbojet cycle), and subsonic cruise range factor almost equal to a supersonic cruise

range factor with the goal that subsonic fuel consumption be at least halfway between those of a turbojet and a bypass ratio 5 turbofan. Currently defined study variable-cycle engines meet these requirements.[171]

A statement of the important advances in propulsion technology attributable to the SCR and VCE programs was recently made by Driver:

Of particular interest has been the evolution of airflow management to reduce off-design penalties. This gain has been made possible, in part, by the advent of digital controls. This technique has also resulted in the term variable-cycle engine, which implies operation at cruise like a turbojet, and operation at off-design conditions similar to a turbofan. Advanced subsonic engines now operate with 6 or 7 internal control variables, whereas these advanced supersonic engines will have more than twice as many control variables — partially as a result of variable inlets and nozzles. The control of off-design performance has made important reductions in the fuel reserve requirement, paying important dividends in additional payload range. It should be recognized that both G.E. and Pratt & Whitney have actually run the critical engine features on a test stand and verified the design features studied in the SCR/VCE program. In the G.E. case, the variable-cycle engine features have been incorporated in a J-101 engine and the performance gains verified on the test stand with a running engine.[172]

Thus, the SCR/VCE program identified two advanced engines that showed promise of efficiently meeting the supersonic aircraft requirements, pointed out several methods that could be used to reduce the engine noise and emissions to acceptable levels, and provided some preliminary validation tests to determine whether all of the elements could be put together in a unit. Much additional effort would have to be spent to make these concepts into real supersonic cruise aircraft engines, but the elements of the technology had been identified.

It should be noted that "variable cycle engines" identified in the NASA SCR and VCE programs do not represent significant advances in propulsion cycle efficiencies over previous supersonic turbofan or turbojet engines. The advantage of these engines is the element of airflow control that provides for a better overall performance on the supersonic cruise mission than either a turbofan or a turbojet engine would provide. In addition, these engines would deliver the improved overall performance with less weight, less fuel, and reduced noise.

SUPERSONIC CRUISE TECHNOLOGY

PROGRESS IN CONFIGURATION CONCEPTS/INTEGRATION

As indicated in Chapter 7, one of the major technical problems of the past SST program was the integration of a myriad of disciplinary concepts into an acceptable SST. In many instances, the aerodynamicist was looking for solutions to aerodynamic problems, the structures engineer was after a solution to structures problems, the propulsion engineer was working on propulsion problems, etc. Actually, all of these disciplinary personnel should have been working on airplane problems because, if the disciplinary solution could not become an integrated solution to the airplane problem, it was really not a solution. Because of this overriding importance of integrated technology in the complicated SST mission, the NASA SCR program made mission performance integration the central element of the program. (See Chapter 8.) The three industry teams (Boeing, Douglas, and Lockheed) and one NASA in-house team not only assessed the applicability of disciplinary concepts, but also introduced disciplinary concepts and integrated concepts.

It is very clear that the progress on the VCE was attributable mainly to the day-to-day interplay between the mission integration or systems studies teams and the propulsion manufacturers. The integration team could rapidly point out where the propulsion system had a shortcoming in thrust or performance, and the propulsion manufacturer could make modifications to overcome this shortcoming. Many iterations were made on the VCEs before they measured up to the SST mission.

From the standpoint of technology growth, it was fortunate that each of the mission integration teams had a different overall configuration concept on which to assess the technology advances. Douglas opted for a highly swept arrow-wing concept similar to the NASA SCAT-15F, but designed for a lower cruise Mach number (2.2 compared to 2.7 for the SCAT-15F). With the lower design Mach number and the noise suppressor technology developed in a joint Douglas/NASA effort, Douglas hoped that the SST mission could be accomplished with a straight turbojet engine or a very low bypass turbofan. Lockheed also chose an arrow-wing configuration, but placed two of the four engines

McDonnell-Douglas reference concept for technology assessment.

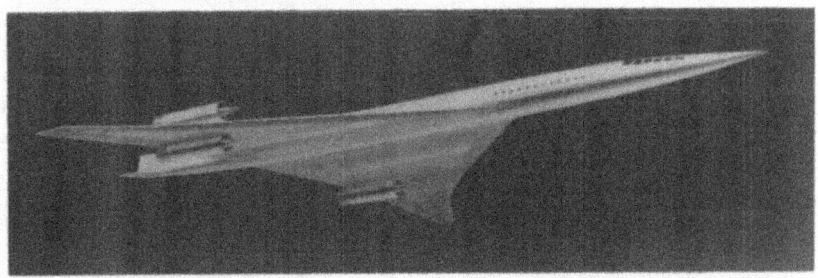

Lockheed reference airplane for SCR technology assessment. Note over/under engine locations on wing.

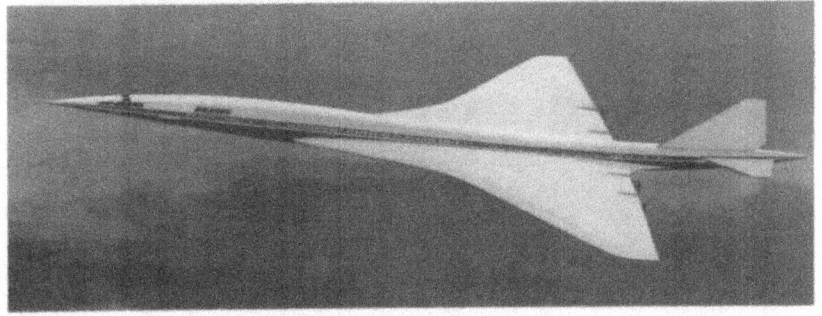

Boeing technology assessment concept for SCR — the improved blended-wing version of Dash 300 SST.

143

unconventionally above the wing for noise shielding purposes. Cruise Mach numbers between 2 and 2.55 were considered. Boeing made use of an updated blended version of the delta-wing 2707–300 airplane that was under study as the U.S. SST candidate. Because of this choice, both Boeing and the SCR program had the benefit of a storehouse of information that had been built on this concept. Boeing considered cruise Mach numbers between 2.2 and 2.7. The final reference configuration was the NASA in-house concept that made use of an updated highly swept SCAT–15F configuration that was proposed to cruise at Mach 2.7. The NASA in-house mission integration team used the modified SCAT–15F as an evaluation concept, and the NASA Lewis Research Center used it in their engine cycle studies early in the SCR program. A large data base was also available on this configuration because of its study by Boeing, Lockheed, and Rockwell International during the U.S. SST program.

Several important disciplinary technology advances were spurred by the mission integration teams. Among these were the Douglas suppressor effort and the Lockheed noise-shielding concepts discussed in the engine noise section, the family concept of SST design introduced by Boeing, and the twin-fuselage idea suggested late in the SCR program.

Supersonic technology demonstrator airplane proposed by Boeing (courtesy of Boeing).

Douglas noise-suppressor models (courtesy of McDonnell-Douglas).

DOUGLAS NOISE SUPPRESSOR CONCEPT

Douglas chose to take a "near term" reference airplane concept with a Mach 2.2 cruise speed and a relatively simple mini-bypass engine. If this airplane was to have a chance to make a good supersonic cruise aircraft, even with advanced aerodynamic and structures technology, some sort of noise suppressor would be required. Consequently, Douglas designed and statically tested a number of small suppressor models. One of these models, a multitube/lobe concept, showed excellent static noise reduction capability. However, many mechanical noise suppressors have demonstrated significant levels of noise reductions at static conditions, but have lost effectiveness at forward velocity. In addition, any significant noise reduction was accompanied by a great loss of thrust,[173] a very undesirable characteristic. For the Douglas suppressor to be different, it would have to be proven before any real technology advance could be presumed. And, as mentioned earlier in this chapter, the Douglas suppressor proved to be different after a concentrated series of wind tunnel, spin rig, and flight tests involving Rolls Royce, the Douglas Company, the NASA SCR Office,

Most promising Douglas suppressor validated in wind tunnel/flight test programs (courtesy of McDonnell-Douglas).

British Aerospace, and the NASA Ames Research Center. The fact that the Douglas suppressor passed the tests[174] was important because the other noise reduction techniques (i.e., coannular effects, noise shielding, and advanced operating procedures) could not in themselves bring the engine noise to desirable levels.

NOISE SHIELDING

The placement of two of the engines above the wing so that the wing and fuselage could be used as a noise shield was an idea proposed by Lockheed early in the SCR program. As pointed out earlier in this chapter, this placement provided a substantial noise reduction of 3 to 5 EPNdB. However, the technology areas that this opened for further study and data accumulation were just as important. The airflow patterns and pressures on the upper surface of a highly swept wing are quite different than those on the lower surface where the engines had

146

been placed in the past. The technical questions about whether the aerodynamic design and analysis programs would handle this unconventional case and how this engine placement would affect inlet design and operation, etc., had to be answered. This engine placement would also probably make more of the wing trailing edge available for controls, and this factor had to be considered. Thus, the desirable goal of noise reduction led to disciplinary problems in aerodynamics and design integration.[175]

AIRPLANE FAMILY CONCEPT

One of the major strengths of the American airplane industry has been the development of families of subsonic jet transports from a given basic design by fuselage "stretching" or small evolutionary changes. These airplane "families" have met the operating requirements of a variety of airlines and route structures, and this approach has led to large savings in costs to the manufacturers and airlines. A big weakness of both the Concorde and U.S. SST programs was that each was essentially based on a "one airplane" effort. Because of the more critical dependence of airplane resistance (drag) on airplane shape (area distribution) in the supersonic flight regime, the practice of fuselage stretching or lengthening did not appear to be feasible. In addition, it was far too costly to develop and certify differently sized airplanes and engines to meet the specific requirements of each of the airlines. As a result of these factors, a "one-sized" Concorde was proposed to cap'ure the North Atlantic passenger traffic from the 120- to 140-passenger Boeing 707 and Douglas DC-8 subsonic transports that were then available. By the time the Concorde was in service, however, the competition on the North Atlantic had grown to include the 250- to 450-passenger "jumbo" jet aircraft, the Boeing 747, the Douglas DC-10, and the Lockheed L-1011. The Concorde could not compete with the productivity of these large aircraft even with a 2.5-times cruise speed capability. A larger Concorde would have preserved the productivity advantages of the SST.

A breakthrough in the supersonic "airplane family" idea was turned in by the Boeing mission integration team during the SCR program.

147

SUPERSONIC CRUISE TECHNOLOGY

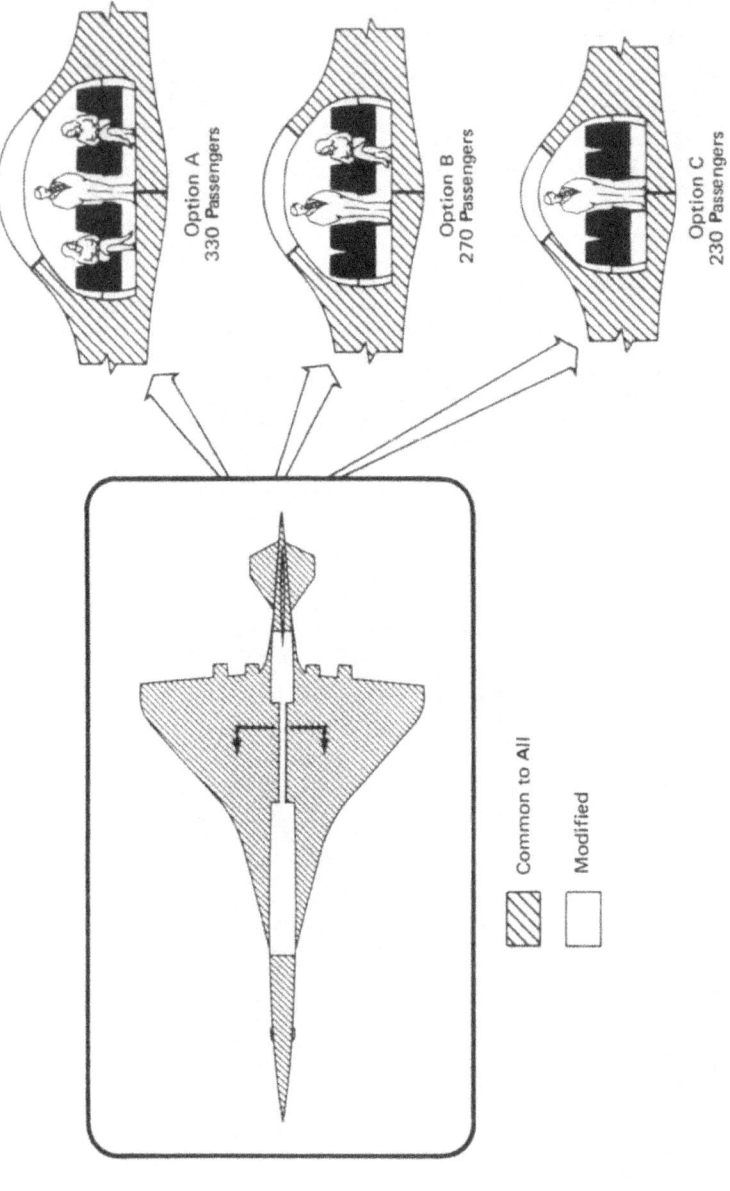

Option A
330 Passengers

Option B
270 Passengers

Option C
230 Passengers

Common to All

Modified

Important airplane family concept for SST airplanes introduced by Boeing (courtesy of Boeing).

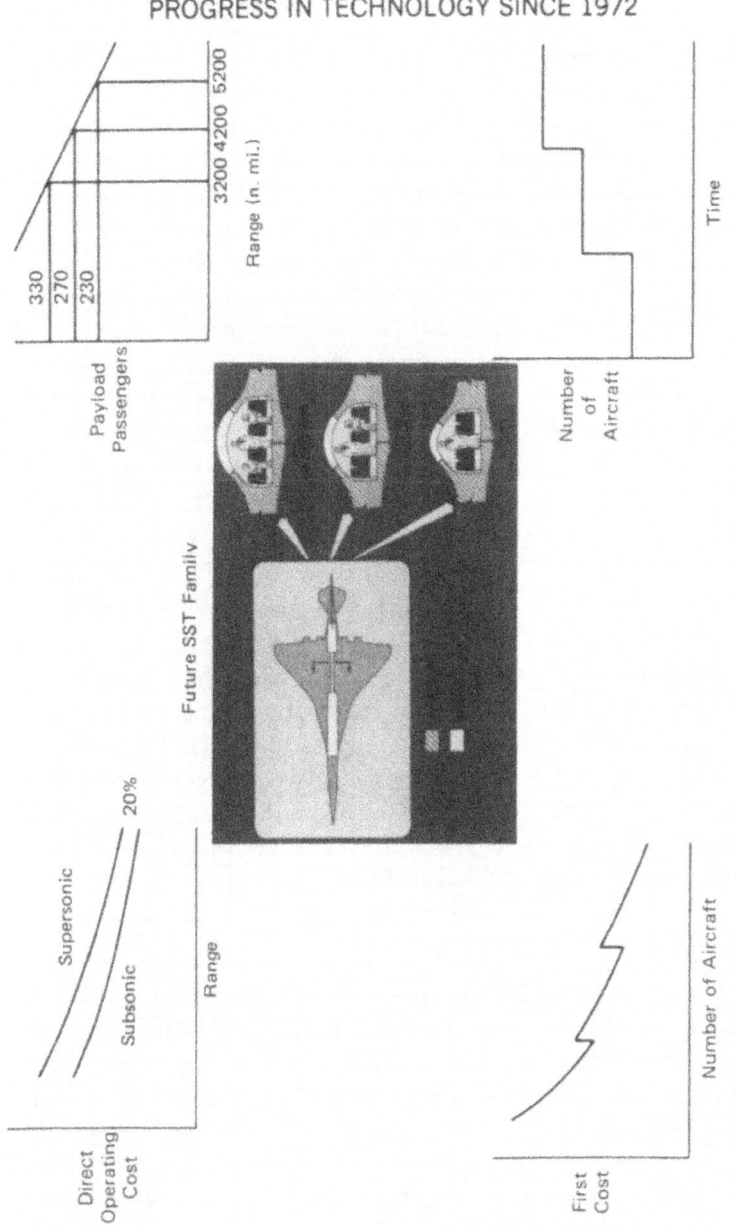

SST family concept provides versatility and cost advantages not present in other SST programs.

149

SUPERSONIC CRUISE TECHNOLOGY

As a means of increasing the marginal cruise efficiency of their delta-wing reference airplane, the Boeing team considered the use of wing/body blending to improve cruise aerodynamics.[176] Blending provided a 16 percent increase in cruise lift/drag,[177] and the study indicated that a supersonic family of airplanes might be possible with the use of lateral fuselage "stretch" rather than the longitudinal stretch that had been employed in creating a subsonic "family." Preliminary results showed that the lateral stretch provided large changes in passenger capacity with relatively small changes in construction and performance when compared to the original design. Before this approach could be adopted, however, aerodynamic and structural validation would have to be accomplished and emergency egress and safety problems would have to be considered. The "proof of concept" studies were completed with the use of Boeing "in-house" and SCR-supported manpower with the result indicating the "lateral-stretch" concept to be feasible for either blended or conventional wing/body structures. The results also suggest that future supersonic airplanes can and should be designed as members of a family.[178]

The Boeing study of the "family" concept indicated that, by means of fairly simple body inserts, supersonic transports can be evolved with payload variations from 230 to 330 passengers and ranges from 3200 to 5200 nautical miles with essentially the same levels of aerodynamic efficiency. This important technology will allow the designer to match various airline payload range requirements while keeping the expensive parts of the airplane unchanged.[179]

LARGE-PAYLOAD SST CONCEPTS

Toward the end of the SCR program, Boeing conducted a study on large-payload supersonic cruise airplanes that could compete with future large double-decked subsonic airplanes. One of these was a large supersonic transport with two decks that could fly from New York to Paris with 500 passengers. Another concept that Boeing considered made use of twin fuselages separated by the engine package. Initial evaluation of this concept indicated the surprising result that perhaps a 60 percent increase in volume could be obtained at no cost in

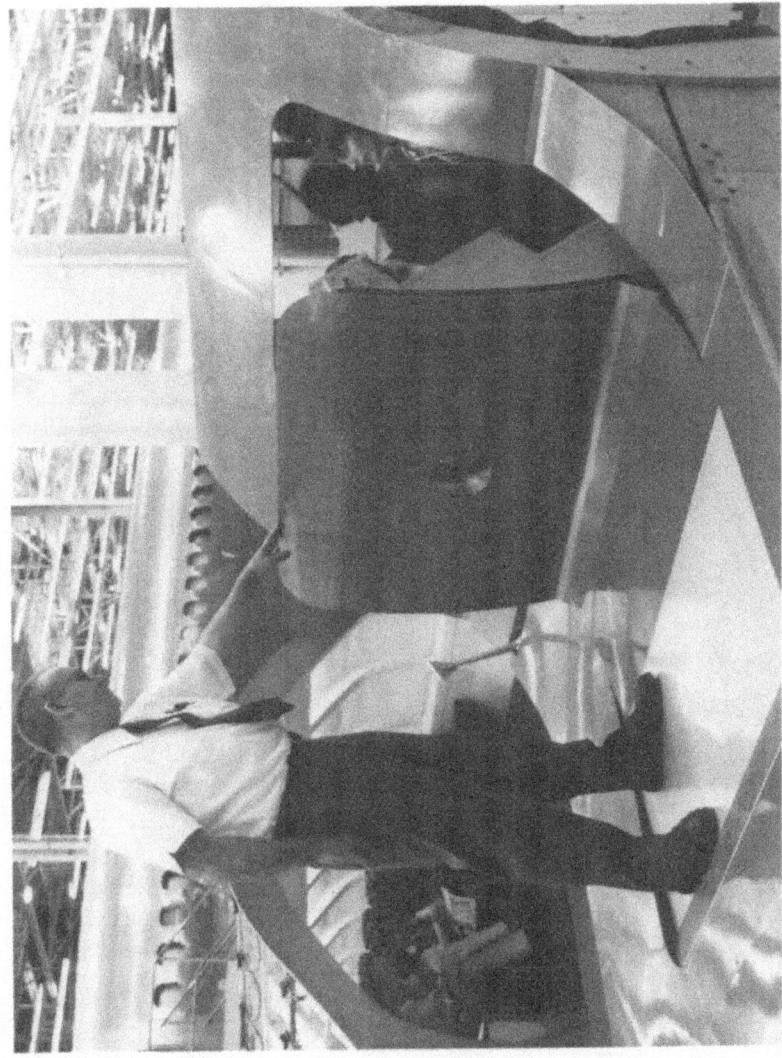

Emergency egress tests of Boeing blended family concept (courtesy of Boeing).

Emergency egress tests continued (photo courtesy of Boeing).

Windowless blended section of Boeing's technology assessment concept (courtesy of Boeing).

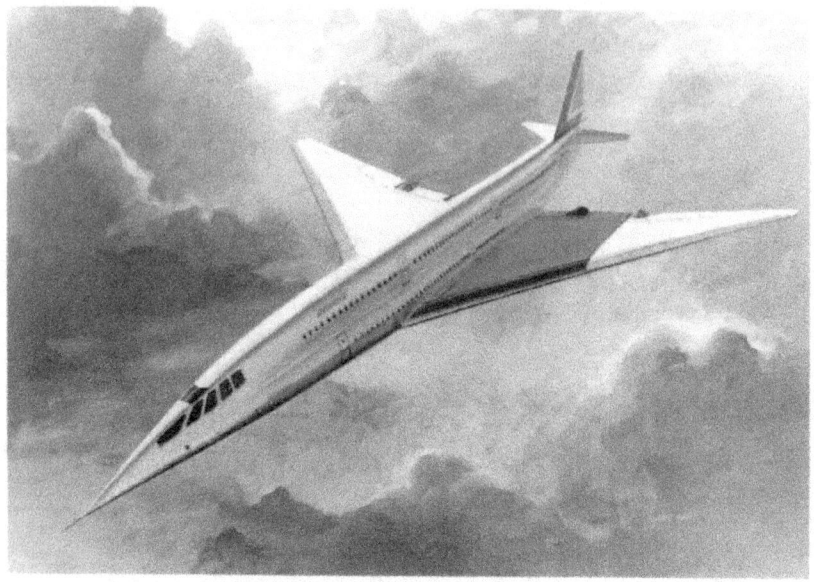

Large-payload, high-productivity concept proposed by Boeing — double-deck passenger facilities (courtesy of Boeing).

aerodynamic efficiency.[180] These preliminary results, although exciting in their meaning to supersonic flight efficiency and productivity, were not validated to the same depth as other elements of the SCR program because of the termination of the program.

OTHER TECHNICAL INNOVATIONS

A number of other important concepts were studied during the SCR program and were brought to a greater state of technology readiness. Among these were the use of active controls to suppress undesirable aeroelastic deflections or vibrations without resorting to heavier, more costly structural elements,[181] and the use of active-control landing gear to reduce the forces on the wing during landing impact.[182] Both of these concepts promised reductions in weight or improvements in fatigue life. Another important concept developed by Boeing was a leading-edge flap that could control the vortex on the leading edge of highly

154

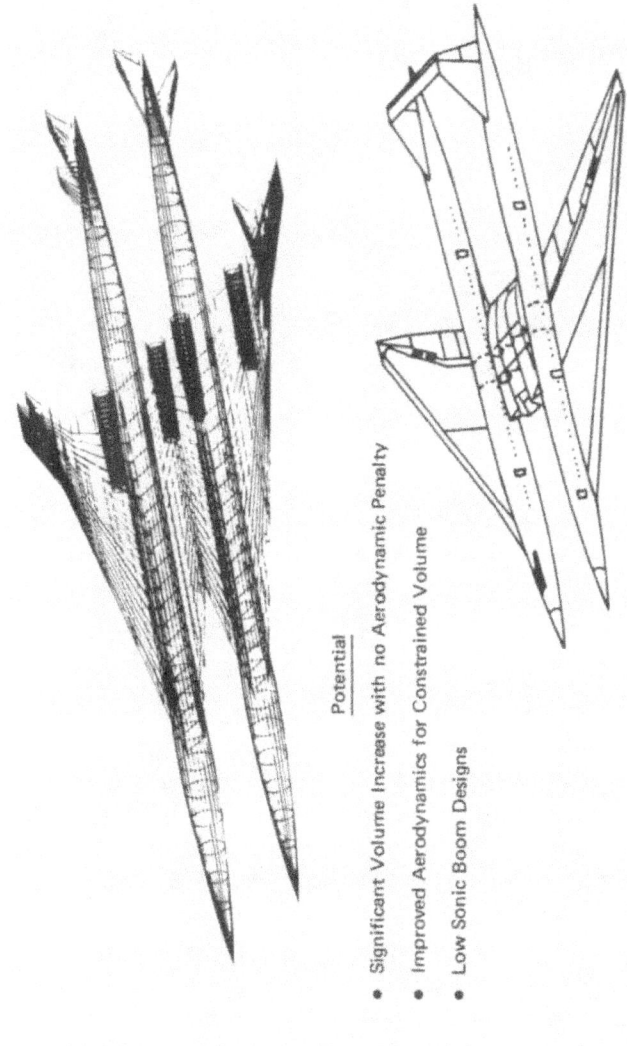

Potential

- Significant Volume Increase with no Aerodynamic Penalty
- Improved Aerodynamics for Constrained Volume
- Low Sonic Boom Designs

Another Boeing large-payload, high-productivity concept — a twin-fuselage SST.

Artist's concept of twin-fuselage SST.

swept arrow wings and prevent flow separation.[183] This flap development was one of the principal elements leading to an improvement in
the low-speed lift/drag of arrow wings, one of the primary goals of the
SCR aerodynamic performance discipline.

SUMMARY OF PROGRESS

During the course of the NASA SCR and VCE programs
(1972–1981), substantial technical progress was made in all elements of
technology relating to supersonic cruise flight, with the possible exception of sonic boom. (As mentioned earlier, the SCR program did not
have sufficient funding to determine the level of acceptability of sonic
boom, and an operating rule was in effect that forbade the supersonic
overland flight of commercial aircraft.) The science of technology integration was also advanced, and a number of new concepts were identified that could lead to acceptable, competitive supersonic transport
aircraft. In the next chapter an attempt is made to quantify the SCR
and VCE progress and to discuss the possible ramifications of this
progress.

CHAPTER 10

Supersonic Cruise Technology Before (1971) and After (1982) the NASA SCR and VCE Programs

Although much supersonic technology progress was made before and during the SST program (see Chapters 3 and 4), the cancellation of that program in 1971 raised many questions about the status of technology. The combined improvements and advances of several decades of research were not sufficient to ensure a viable supersonic transport. Both the Concorde and the proposed U.S. SST were marginal from the standpoint of aerodynamic, propulsive, and structural efficiencies, and both were projected to have critical environmental problems in the areas of engine noise, pollution, and sonic boom. No technology for solving these problems appeared to be on the horizon, and as a result, the U.S. SST program was canceled and the Concorde production program was severely curtailed.

Over the decade since the cancellation of the U.S. SST program, the NASA SCR and VCE programs have had the responsibility of filling this supersonic technology void and providing the base for a viable supersonic transport. Some of the progress from this technology effort was described in the previous chapter. This chapter will present quantitative comparisons of the technology progress and will discuss some of the possible ramifications of the results.

COMPARISONS

EMISSIONS

Most of the progress on engine emissions was made outside the SCR and VCE programs, but it was still principally motivated by the questions that had arisen about the supersonic transport. The NASA Lewis Research Center effort on burner technology provided practical burners for SCR application that had nitrous oxide emission indices of about one-half of those that existed at the end of the U.S. SST program. Conceptual burners were considered that can lead to emission indices less than 20 percent of conventional burners.

NASA participated in studies for improving the understanding of the effect of engine emissions on the depletion of the protective ozone layer in the atmosphere. Although this influence was largely unknown at the close of the SST program, the DOT CIAP and the HAPP programs have done much to answer the questions. Where the earlier CIAP and National Academy of Sciences studies showed a substantial, but non-critical, depletion of ozone due to engine emissions, later HAPP studies have shown that this effect might actually lead to a small increase in ozone. This represents a rather dramatic reversal in the technical understanding of the atmosphere, but it does not alter the fact that this question should be a continuous subject of study.

ENGINE NOISE

As mentioned earlier, the only noise reduction technology available at the end of the U.S. SST program was the use of large oversized engines that resulted in extremely large performance losses by the proposed SST. These performance losses would prohibit the SST from making a minimal North Atlantic mission. The SCR and VCE programs substantially improved this technical situation by identifying and partially validating four methods of noise reduction: mechanical suppression, noise shielding, coannular effects, and advanced operating procedures. It is estimated that a combination of these effects could be used on a current SST design to reduce the noise 10 to 12

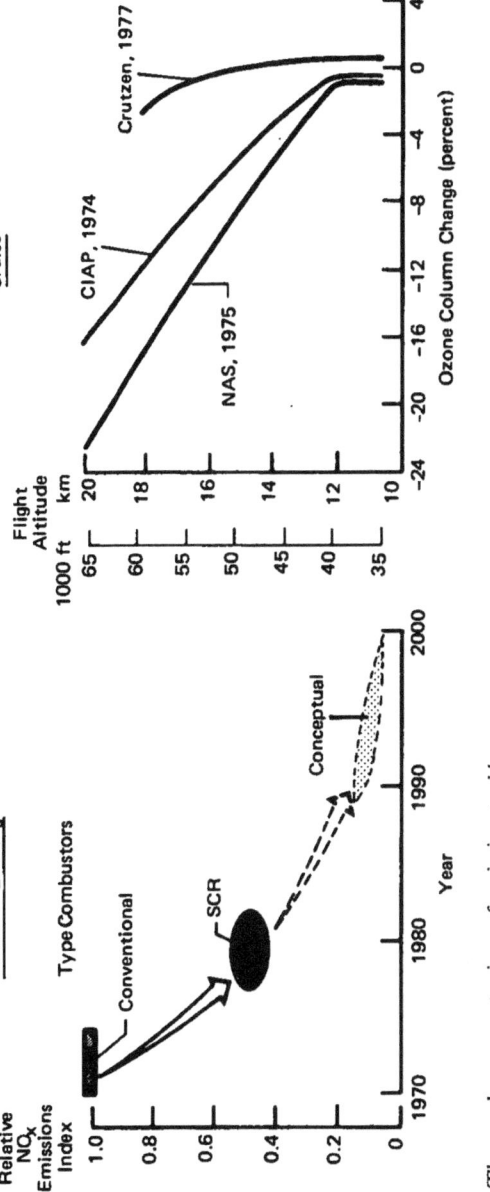

Then and now comparison of emission problems.

159

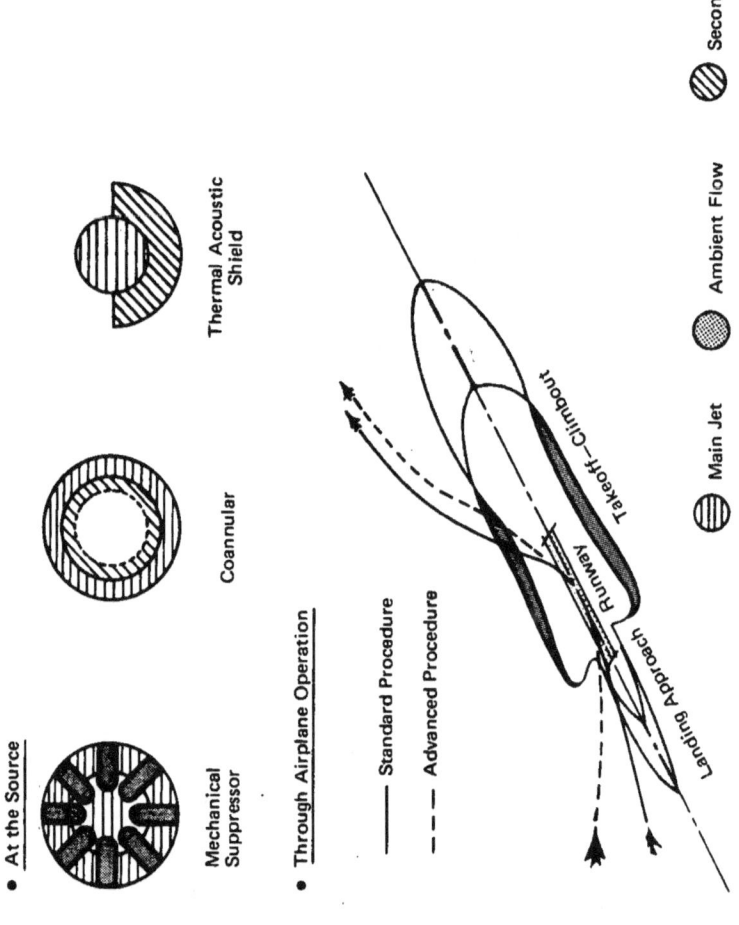

Principal SCR/VCE noise-reduction methods.

decibels (dB) below the levels expected of the U.S. SST concept. Another 10-dB reduction would likely result from a continuing focused research program on supersonic cruise aircraft over the next 8 to 10 years.

These identified noise reductions are quite dramatic and indicate that the noise exposure levels for supersonic cruise vehicles can be comparable to their equivalent subsonic counterparts.[184]

AERODYNAMICS

At the close of the U.S. SST program in 1971, the cruise aerodynamic efficiency of the proposed airplane concept was marginal, and no means were readily available for improving this efficiency. Concepts such as the highly swept arrow wing provided the desirable cruise efficiency, but were aerodynamically deficient at low speeds. This technical quandry has been somewhat alleviated by aerodynamic research conducted by NASA and SCR contractors. Through the use of wing/body blending, along with a slight strake extension, the cruise aerodynamic efficiency of the delta-like SST configuration has been improved by about 16 percent to come within 10 percent of the highly swept arrow wing and 33 percent better than the Concorde. In addition, the SCR aerodynamics program identified flap arrangements that produce a 16 percent improvement in the low-speed lift/drag of the arrow wing compared with 1971 levels when plain flaps were used. These aerodynamic improvements have very favorable connotations for both concepts.

STRUCTURES AND MATERIALS

As the U.S. SST program drew to a close in 1971, the proposed airplane had an aeroelastic problem of serious magnitude, and this problem was further compounded by a late decision to consider oversized engines for reducing noise. The proposed titanium structure was acceptable, but the fabrication costs were expected to be high. Along with improved methods for designing and assessing aeroelastic characteristics, the NASA SCR program has provided some structural

SUPERSONIC CRUISE TECHNOLOGY

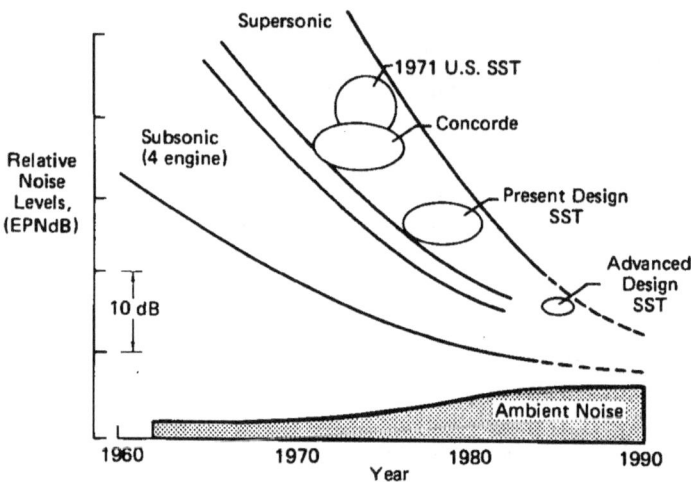

Then and now noise status.

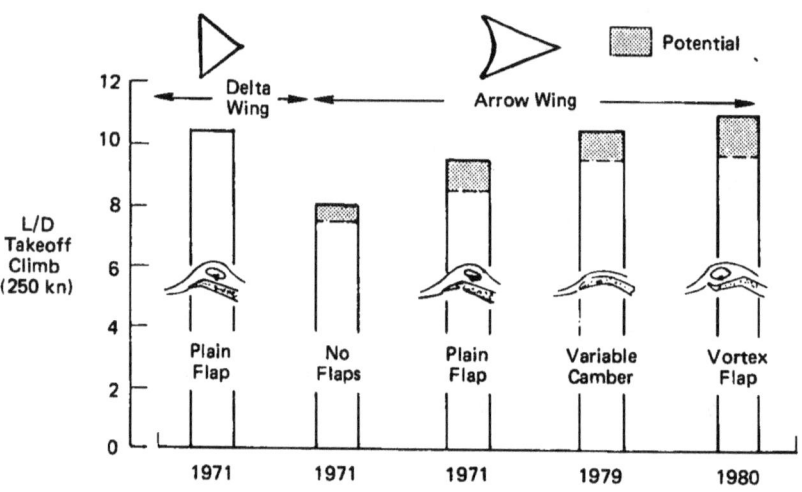

Then and now comparison of low-speed aerodynamic efficiency.

162

BEFORE/AFTER SCR AND VCE PROGRAMS

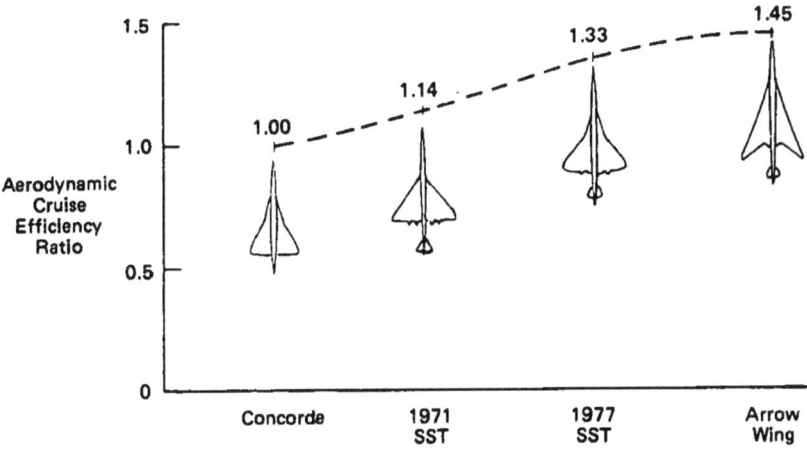

Then and now comparison of supersonic cruise efficiency.

concepts and fabrication processes that offer drastic reductions in the weight and/or cost of supersonic structures. The superplastic forming and concurrent diffusion bonding (SPF/DB) of titanium, first proposed by Rockwell International and then brought to a high state of technology readiness by Douglas, offers a 25 percent reduction in weight and a 33 percent reduction in cost when compared to conventional titanium plate. Further improvements developed by Douglas include the use of fiber-reinforced advanced titanium (FRAT) to provide additional reductions in weight at some cost penalty. Weight reduction is one of the most powerful means for improving the performance of an airplane.

PROPULSION

The General Electric GE-4 afterburning turbojet engine that was proposed for the U.S. SST was perhaps the best compromise that technology could provide in 1971. It provided reasonably good supersonic performance and, with the use of afterburners, provided adequate thrust for takeoff and acceleration to cruise speed. Because it was basically a turbojet engine, however, its performance at subsonic

163

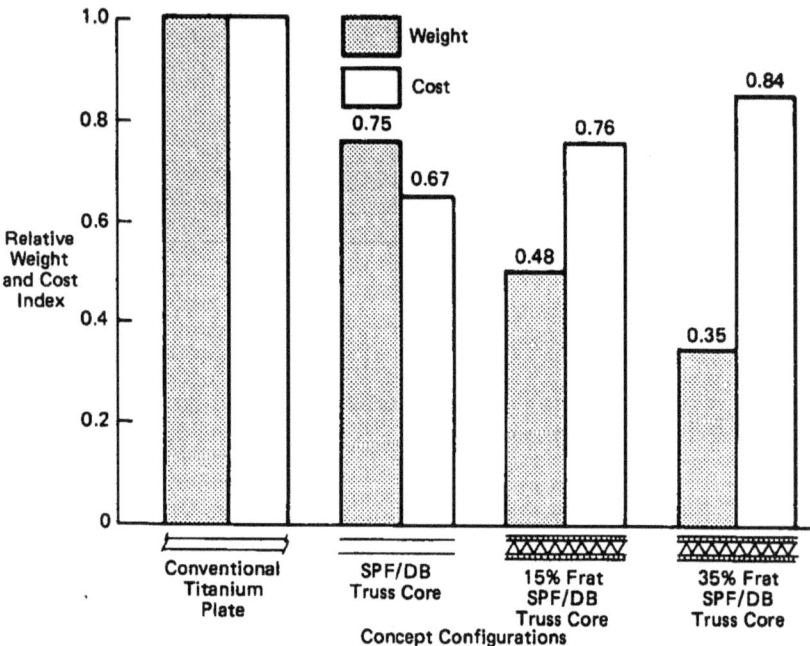

SUPERSONIC CRUISE TECHNOLOGY

Benefits from SCR advanced structures and materials program.

speeds was only nominal. In addition, the use of the afterburner for takeoff and climb implied noise levels that were well above the 108-EPNdB level established by the FAR 36 noise rule placed in effect in 1969. The proposal to use an oversized GE-4 to reduce noise not only created drastic problems in airplane integration, but would also have led to performance degradations in the propulsion system.

Perhaps the most spectacular development of the SCR and VCE programs was the replacement of the uncertain 1971 propulsion technology with the "variable cycle engine" and the component technologies for reduced noise and emissions. When compared to the 1971 GE-4, the VCE provides a 10 percent reduction in supersonic cruise specific-fuel-consumption (sfc), a similar reduction in transonic sfc, and a remarkable 24 percent reduction in the sfc at important subsonic speeds. Furthermore, the VCE provides this improved perform-

164

ance with a weight that is only 75 percent of that of the GE-4. This represents a tremendous gain in propulsion technology, and remember, many elements of the VCE were validated during the SCR and VCE programs.

MISSION INTEGRATION

It is difficult to quantify the technology improvements in mission integration or the impact of the SCR concepts on technology for supersonic cruise flight. It is certain, however, that these systems integration studies were responsible for many of the disciplinary technology advances, and concepts such as the "supersonic airplane family" offered solutions to long-time questions.

POSSIBLE RAMIFICATIONS OF SCR AND VCE TECHNOLOGY

With no current effort in the United States involving the area of supersonic cruise flight, there are no realistic projections for the use of

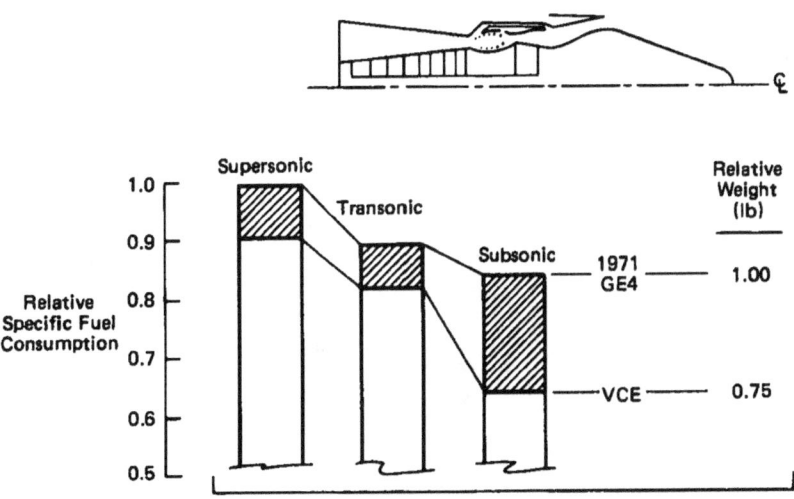

Improvements in propulsion as a result of SCR/VCE programs.

165

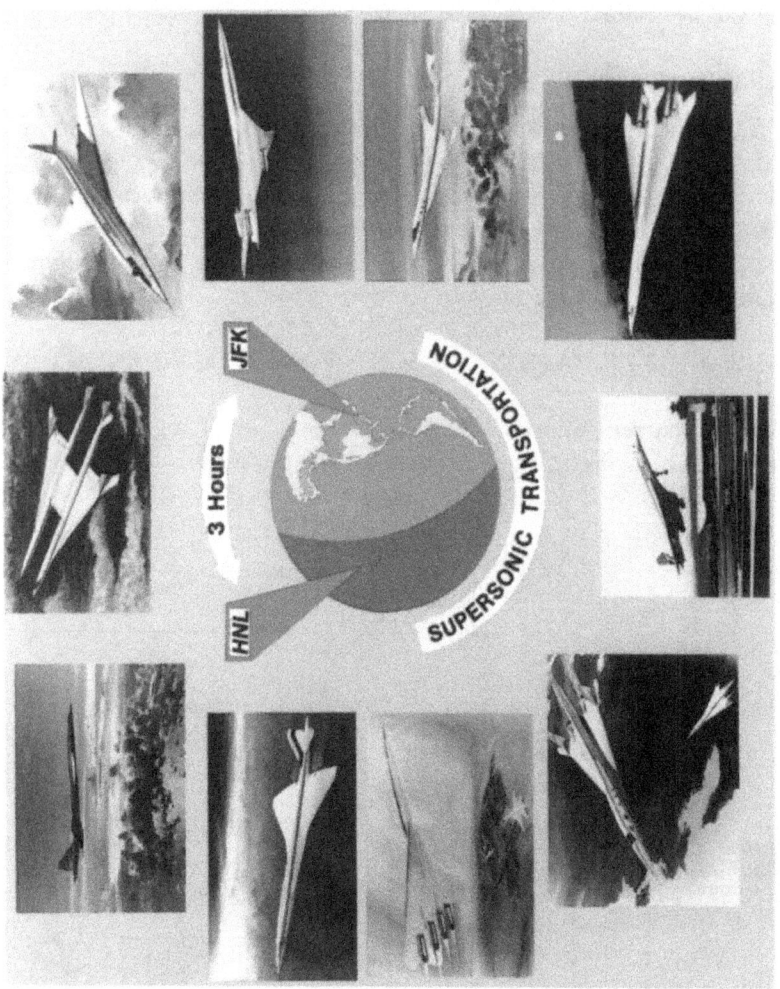

Airplane concepts for supersonic transportation.

SCR or VCE technology other than the isolated elements that will find their way onto subsonic transports or on short-range supersonic military aircraft. There is no military interest in supersonic cruise flight in the true sense, and there appears to be little interest in this concept of flight within the U.S. Government or the airplane industry. There are sporadic bursts of activity on the possibility of using SCR and VCE technology to develop a supersonic business jet transport,[185] but there is no way to assess the sincerity of these efforts. Certainly a large amount of money would be required to develop such an airplane, and it would almost assuredly have to come from private sources.

Although there are no present potential applications of the SCR and VCE technology efforts, there are still possibilities for the development of an outstanding supersonic transport. As mentioned earlier in this chapter, the noise and emissions would be greatly reduced from prior SST levels, and the technology advances in structures, aerodynamics, and propulsion would lead to great improvements in range/payload capability, fuel use, cost, and economics.

RANGE/PAYLOAD CAPABILITY

If we forget for a minute that the 1971 SST had perhaps unacceptable noise characteristics, it was to transport 280 passengers a distance of 3600 nautical miles. With the use of SCR and VCE "airplane family" technology, approximately 385 passengers could be transported the 3600-nautical-mile distance, a payload improvement of 37.5 percent. Or, the 280-passenger payload of the 1971 SST could be transported a distance of 4950 nautical miles with SCR and VCE technology, a range improvement of 37.5 percent. In providing this 37.5 percent improvement in range/payload capability, the SCR/VCE technology SST would burn only about one-half as much fuel per seat-mile as the 1971 SST.

ECONOMICS/PRODUCTIVITY

One means of measuring the economics of supersonic transports is to determine the ticket surcharge that the supersonic passenger would

167

SUPERSONIC CRUISE TECHNOLOGY

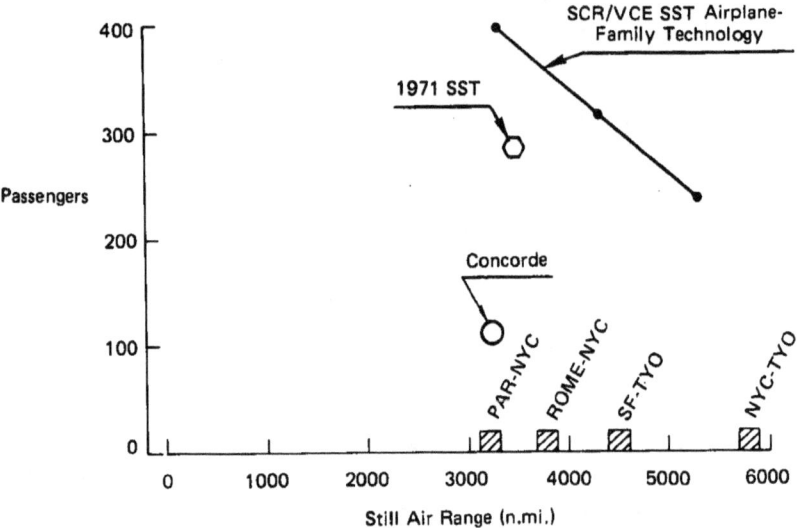

Range/payload capability then (1971) and now (1982).

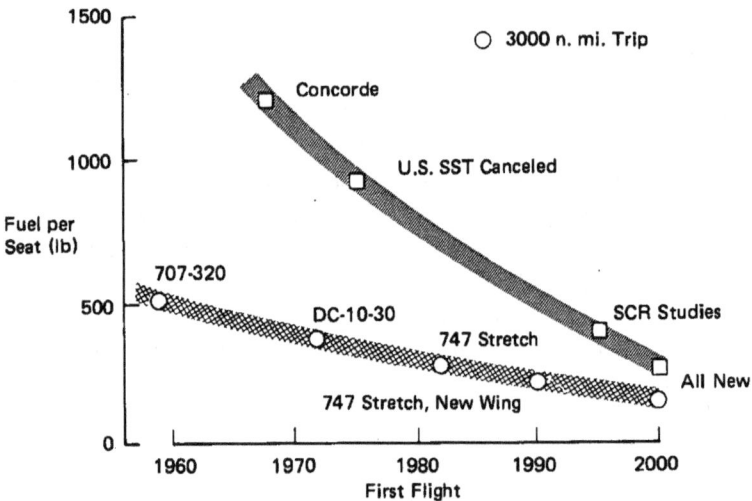

Fuel-use trends as a result of SCR/VCE programs.

have to pay for the airline to make a return on investment equal to the subsonic transport. For the 1971 SST, the required surcharge was about 50 percent above the subsonic tourist rate. The SCR and VCE technology would provide the basis to drive this surcharge to nearly zero.[186] The major reason for this, of course, is the fact that the SCR and VCE programs have been able to restore speed to the productivity equation at very little increase in total operating costs.

OTHER POSSIBLE RAMIFICATIONS

The NASA SCR and VCE technology could lead to supersonic military-troop and equipment-moving transports that would give real meaning to the term "rapid deployment forces." There is also some possible application of the technology to the development of a supersonic missile carrier.[187]

Now that the NASA SCR and VCE programs have identified solutions to some of the major problems of supersonic cruise flight, what is the future course of the effort? The final chapter addresses this question.

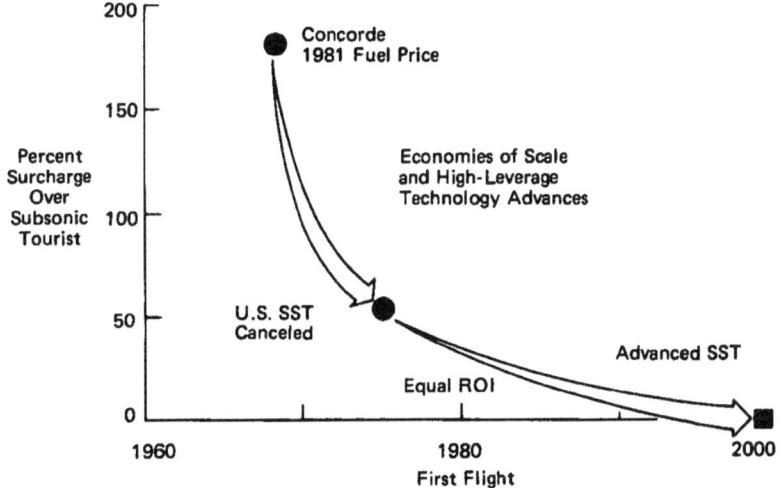

Required ticket surcharge for supersonic travel.

SUPERSONIC CRUISE TECHNOLOGY

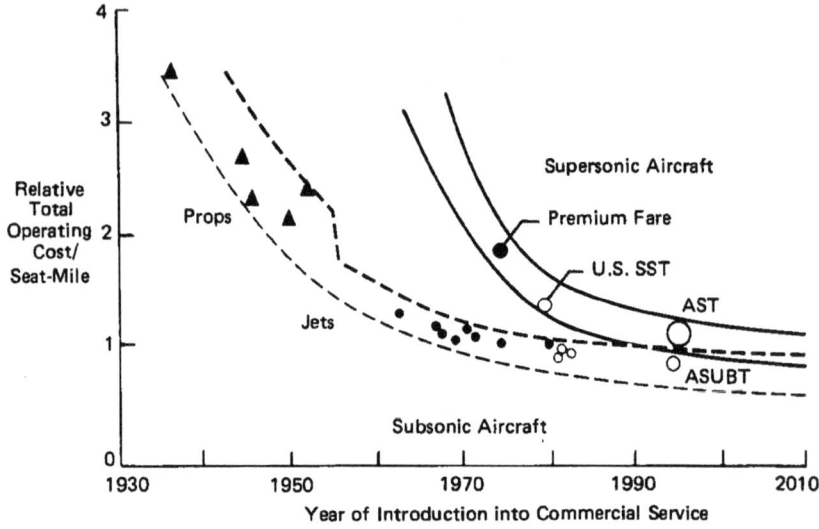

Projected operating cost of supersonic transport approaches that of advanced subsonics.

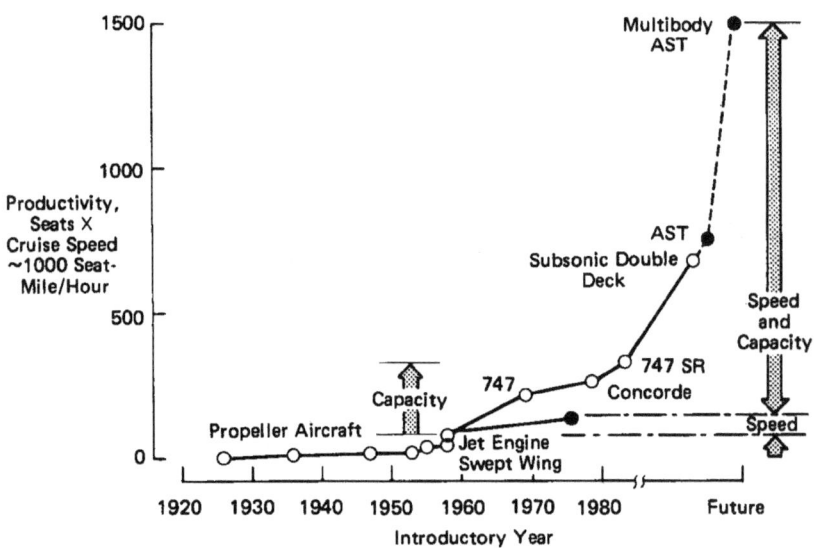

Productivity history and projection for the long-range airline market.

170

CHAPTER 11

Future Directions of Supersonic Cruise Research

During the past decade, the NASA SCR and VCE programs have identified technology that will meet many of the needs of future supersonic cruise aircraft. At a cost of about 7.5 cents per year for each American, these programs have identified and validated many elements of an important new engine concept—the variable cycle engine; identified and partially validated four promising means of noise reduction; built and tested a wide variety of wind tunnel models to establish a comprehensive aerodynamic data base; promoted and spread the technology of superplastic forming and bonding of titanium from which grew the FRAT effort; answered the age-old problem of SST sizing with the SST airplane family concept; developed two large-payload, high-productivity concepts that can compete with almost any future airplane and may, because of their size, have sonic boom advantages; and promoted many improvements to the analytic methods for structural design and analysis. Quite important also, the SCR and VCE programs have kept a cadre of skilled, dedicated supersonic specialists alive in the airplane industry. These teams, along with NASA and other SCR and VCE contractors, were to continue to validate the elements of supersonic cruise technology that had been identified and continue highly focused programs for identifying even more advanced technology. There were still "bugs" in the SPF/DB and FRAT structural forming processes, and there were some areas of uncertainty in almost every disciplinary field, as well as in the area of concept integration. There were also fertile fields like sonic boom and laminar flow at supersonic speeds that had to be reopened because of the high potential payoff. And, of course, the focused nature of the program would bring further "breakthroughs" that could not be anticipated.

When this document was proposed in early 1981, the final chapter was to outline the future course and milestones of the SCR and VCE programs. Work had been initiated on the critical inlet configurations that were to "feed" air to the variable cycle engines. In addition, wind tunnel models were planned for determining the aerodynamic characteristics of the large-payload twin-fuselage concept that was under consideration, and the focused SCR and VCE programs were to move forward in other areas. However, as a result of budget problems in fiscal year 1982, NASA canceled the focused SCR and VCE programs. With the Soviet SST program in a stagnant state and no further Concordes in sight, there appeared to be no clear and immediate need for SCR technology and there was no clear expression of commercial interest in the VCE program.

The cancellation of the NASA SCR and VCE programs has led to the breakup of the industry teams that participated in the supersonic cruise effort and has left the future status of supersonic cruise technology in doubt. The NASA Lewis Research Center will probably test the inlet hardware that has been prepared, and NASA will continue an effort in supersonic technology. However, this effort will not provide a unified research attack on problems unique to supersonic aircraft and will not generate the interest and support of industry that a focused program would.

FUTURE DIRECTIONS OF RESEARCH

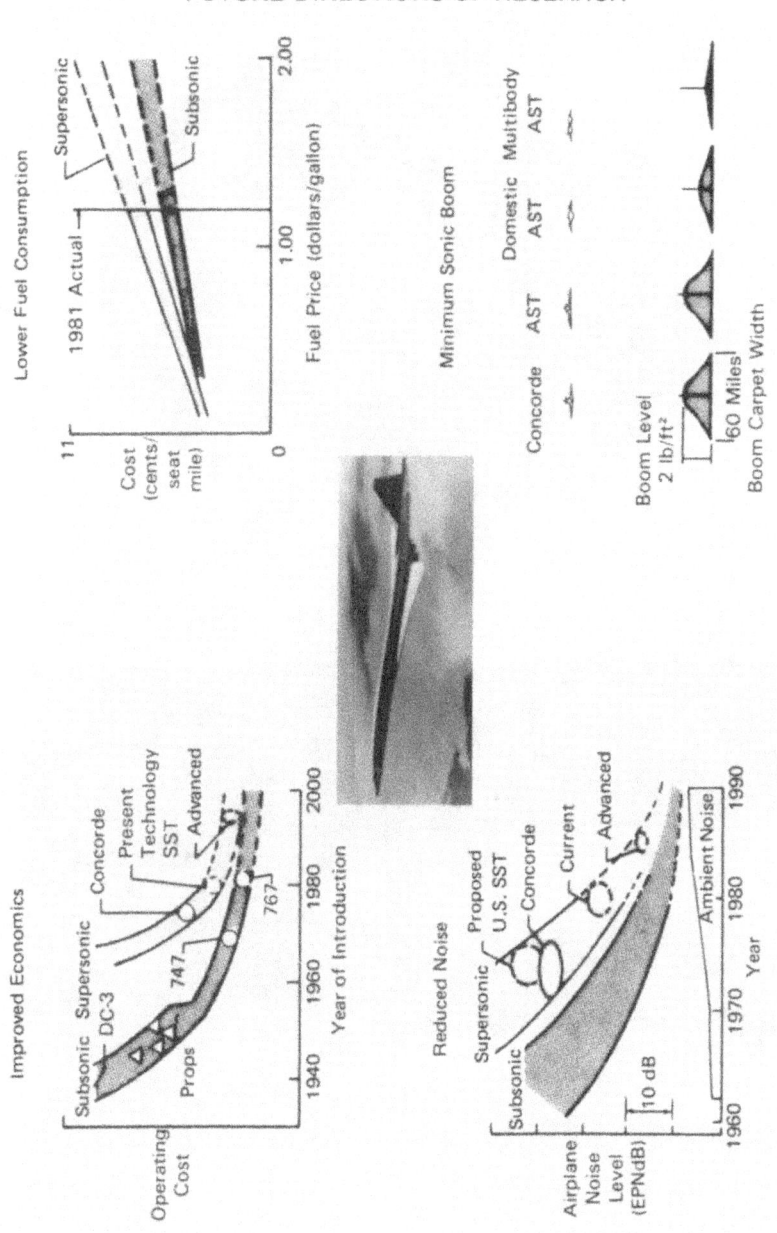

Supersonic transport major needs.

Important SCR/VCE inlet tests will be continued (courtesy of Boeing).

Douglas inlet hardware for tests at NASA Lewis Research Center (courtesy of McDonnell-Douglas).

References

1. National Aeronautics and Space Act of 1958, Public Law 85-568, 85th Congress, H.R. 12575, July 29, 1958, p. 1.
2. Hallion, R., *Supersonic Flight*, The Macmillan Company, New York, 1972, p. 2.
3. *Ibid*, p. 1.
4. Dwiggins, D., *The SST: Here It Comes Ready or Not*, Doubleday & Company, Inc., Garden City, New York, 1968, p. 110.
5. *Ibid.*, p. 110.
6. *Ibid.*, p. 111.
7. *Ibid.*, p. 111-112.
8. *The Supersonic Transport—A Technical Summary*, NASA TM D-423, June, 1960, p. 1.
9. *Commercial Supersonic Transport Aircraft Report*, DOD/NASA/FAA, Washington, D.C., December 1960, p. 2.
10. Baals, D. D., and Corliss, W. R., *Wind Tunnels of NASA*, NASA SP-440, 1981.
11. Technical Facilities Catalog, NASA NHB 8800.5A (I), October 1974.
12. Dwiggins, D., *op. cit.*, pp. 4-5.
13. Supersonic Transport Development Program, FAA, Washington, D.C., June 19, 1963.
14. Dwiggins, D., *op. cit.*, p. 131.
15. *Ibid.*, illustration 33.
16. Costello, J., and Hughes, T., *The Concorde Conspiracy*, Charles Scribner's Sons, New York, 1976, pp. 147-151.
17. *Ibid.*, illustration opposite p. 154.
18. Knight, G., *Concorde: The Inside Story*, Stein and Day, New York, 1976, p. 64.
19. Costello, J., and Hughes, T., *op. cit.*, p. 188.
20. *Ibid.*, p. 189.
21. *Ibid.*, p. 189.
22. *Ibid.*, p. 225.
23. Driver, C., Progress in Supersonic Cruise Technology, *AIAA Paper 81-1687*, August 1981, p. 1.
24. *Ibid.*, p. 5.

25. *Ibid.*, p. 5.
26. Hallion, R., *op. cit.*, p. 13.
27. *Ibid.*, pp. 17–18.
28. *Ibid.*, p. 18.
29. *Ibid.*, p. 21.
30. *Ibid.*, p. 22.
31. *Ibid.*, p. 22.
32. *Ibid.*, p. 12.
33. *Ibid.*, pp. 19–20.
34. *Ibid.*, pp. 20–21.
35. *Ibid.*, pp. 16–17.
36. *Ibid.*, pp. 22–23.
37. *Ibid.*, pp. 24–25.
38. *Ibid.*, p. 23.
39. *Ibid.*, p. 23.
40. *Ibid.*, p. 24.
41. *Ibid.*, pp. 26–27.
42. *Ibid.*, p. 39.
43. *Ibid.*, p. 42.
44. *Ibid.*, p. 43.
45. *Ibid.*, p. 36.
46. *Ibid.*, p. 42.
47. *Ibid.*, p. 109.
48. *Ibid.*, p. 109.
49. *Ibid.*, p. 45.
50. Baals, D. D., and Corliss, W. R., *op. cit.*, pp. 61–62.
51. *Ibid.*, pp. 62–63.
52. *Ibid.*, pp. 65–66.
53. Dwiggins, D., *op. cit.*, p. 34.
54. *Ibid.*, p. 109.
55. *Ibid.*, p. 109.
56. *Ibid.*, pp. 109–110.
57. *Ibid.*, p. 110.
58. Eggers, A. J., and Syvertson, C. A., Aircraft Configurations Developing High Lift-Drag Ratios at High Supersonic Speeds, NACA RM–A55I05, 1955.
59. Dwiggins, D., *op. cit.*, p. 108.
60. *Ibid.*, pp. 110–111.
61. The Supersonic Transport — A Technical Summary, *op. cit.*, p. 1.
62. *Ibid.*, p. 1.
63. Commercial Supersonic Transport Aircraft Report, DOD/NASA/FAA, Washington, D.C., June 1961, p. ii.
64. Dwiggins, D., *op. cit.*, p. 119.

REFERENCES

65. *Proc. NASA Conference on Supersonic-Transport Feasibility Studies and Supporting Research*, NASA TM X-905, 1963.
66. *Supersonic Transport Development Program, op. cit.*, p. 56.
67. Hayes, W. D., Linearized Supersonic Flow, North American Aviation, Inc., Report AL-222, June 18, 1947.
68. Jones, R. T., Estimated Lift-Drag Ratios at Supersonic Speed, NACA TN 1350, July 1947.
69. Whitman, G. B., The Flow Pattern of a Supersonic Projectile, *Commun. Pure Appl. Math.*, Vol. V, No. 3, Aug. 1952, pp. 301-348.
70. Hayes, W. D., *op.* cit.
71. Hayes, W. D., Haefeli, R. C., and Kulsrud, H. E., Sonic Boom Propagation in a Stratified Atmosphere, with Computer Program, NASA CR-1299, 1969.
72. Dwiggins, D., *op. cit.*, p. 189.
73. *Ibid.*, p. 188.
74. Office of Technology Assessment, Impact of Advanced Air Transport Technology, Part 1: Advanced High-Speed Aircraft, OTA-T-112, April 1980, p. 31.
75. *Ibid.*, p. 31.
76. Dwiggins, D., *op. cit.*, p. 213.
77. *Ibid.*, p. 213.
78. *Anon.*, TU-144: 1965-1978, *Flight International*, Sept. 30, 1978, p. 1230.
79. *Ibid.*, pp. 1229-1230.
80. *Ibid.*, p. 1230.
81. *Ibid.*, p. 1230.
82. *Ibid.*, p. 1230.
83. *Anon., Aviation Daily*, Vol. 207, No. 36, June 20, 1973, p. 281.
84. Dwiggins, D., *op. cit.*, pp. 199-200.
85. Costello, J., and Hughes, T., *op. cit.*, p. 25.
86. *Ibid.*, pp. 26-27.
87. *Ibid.*, pp. 27-28.
88. *Ibid.*, p. 39.
89. *Ibid.*, pp. 51-52.
90. Dwiggins, D., *op. cit.*, p. 202.
91. *Ibid.*, pp. 203-204.
92. Costello, J., and Hughes, T., *op. cit.*, p. 97.
93. Office of Technology Assessment, *op. cit.*, p. 27.
94. *Ibid.*, pp. 27-28.
95. Gillman, P., Supersonic Bust: The Story of the Concorde, *Atlantic*, Vol. 239, Jan. 1977, pp. 72-81.
96. Melville, F., The Concorde's Disastrous Economics, *Fortune*, Jan. 30, 1978, p. 67.
97. Leyman, C. S., Concorde with the Airlines, *Supersonic Cruise Research '79*, NASA CP-2108, Part 2, 1980, pp. 714-757.
98. Office of Technology Assessment, *op. cit.*, p. 28.

99. Knight, G., *op. cit.*, p. 167.
100. Dwiggins, D., *op. cit.*, p. 148.
101. *Ibid.*, p. 224.
102. *Ibid.*, p. 26.
103. *Ibid.*, p. 34.
104. *Ibid.*, p. 33.
105. *Ibid.*, p. 33.
106. Costello, J., and Hughes, T., *op. cit.*, p. 144.
107. Hallion, R., *op. cit.*, pp. 15–16.
108. Office of Technology Assessment, *op. cit.*, p. 31.
109. Lynch, D. M., *Barriers to Innovation in Transportation: A Case Study of the SST,* Massachusetts Institute of Technology, 1975, p. 83.
110. *Ibid.*, p. 89.
111. Dwiggins, D., *op. cit.*, p. xv.
112. *Ibid.*, p. xv.
113. Lynch, D. M., *op. cit.*, p. 86.
114. Costello, J., and Hughes, T., *op. cit.*, p. 278.
115. Dwiggins, D., *op. cit.*, p. 163.
116. Lynch, D. M., *op. cit.*, p. 75.
117. *Ibid.*, p. 75–76.
118. *Ibid.*, p. 76.
119. Carlson, H. W., and McLean, F. E., The Sonic Boom, *International Science and Technology*, July 1966, p. 70.
120. Office of Technology Assessment, *op. cit.*, p. 88.
121. Lynch, D. M., *op. cit.*, p. 51.
122. McLean, F. E., Some Nonasymtotic Effects on the Sonic Boom of Large Airplanes, NASA TN D-2877, June 1965.
123. Office of Technology Assessment, *op. cit.*, p. 89.
124. Grobecker, A. J., Coroniti, S. C., and Cannon, R. H., Jr., The Effects of Stratospheric Pollution by Aircraft, DOT-TST-75-50, December 1974.
125. Broderick, A. J., Stratospheric Effects from Aviation, *AIAA Paper 77-799*, July 1977.
126. Office of Technology Assessment, *op. cit.*, p. 90.
127. Lynch, D. M., *op. cit.*, p. 60.
128. Magruder, W. M., *The SST and the National Interest* (SST White Paper), Department of Transportation, Feb. 1971.
129. *Ibid.*, figure G-10-300-7/10/70.
130. Office of Technology Assessment, *op. cit.*, p. 3.
131. Hoffman, S., and Varholic, M. C., Contracts, Grants and Funding Summary of Supersonic Cruise Research and Variable-Cycle Engine Technology Programs, 1972-1982, NASA TM-85650, 1983.

REFERENCES

132. Hoffman, S., Bibliography of Supersonic Cruise Aircraft Research (SCAR) Program from 1972 to Mid-1977, NASA RP-1003, 1977.

133. Hoffman, S., Bibliography of Supersonic Cruise Research (SCR) Program from 1977 to Mid-1980, NASA RP-1063, 1980.

134. Sigalla, A., Overview of Boeing Supersonic Transport Efforts—1971-1979, *Supersonic Cruise Research '79*, NASA CP-2108, Part 2, 1980, pp. 826-827.

135. Duerr, R. A., and Diehl, L. A., Advanced Technology for Controlling Pollutant Emissions from Supersonic Cruise Aircraft, *Supersonic Cruise Research '79*, NASA CP-2108, Part 1, pp. 535-549.

136. Driver, C., *op. cit.*

137. Sundararaman, N., Environmental Effects of Aircraft at Cruise: An Update, *Supersonic Cruise Research '79*, NASA CP-2108, Part 1, 1980, p. 523.

138. Duerr, R. A., and Diehl, L. A., *op. cit.*

139. Driver, C., *op. cit.*, p. 3.

140. *Ibid.*, p. 3.

141. *Ibid.*, p. 3.

142. FitzSimmons, R. D., McKinnon, R. A., and Johnson, E. S., Flight and Tunnel Test Results of the MDC Mechanical Jet Noise Suppressor Nozzle, *Supersonic Cruise Research '79*, NASA CP-2108, Part 1, 1980, p. 453.

143. Wright, B. R., Sedgwick, T. A., and Urie, D. M., An Advanced Concept that Promises Ecological and Economic Viability, *Proc. SCAR Conference*, NASA CP-001, Part 2, 1976, p. 946.

144. Driver, C., *op. cit.*, p. 3.

145. Driver, C., and Maglieri, D. J., Some Unique Characteristics of Supersonic Cruise Vehicles and Their Effect on Airport Community Noise, *AIAA Paper 80-0859*, May 1980.

146. Albers, J. A., and Olinger, F. V., YF-12 Propulsion Research Program and Results, *Proc. SCAR Conference*, NASA CP-001, Part 1, 1976, pp. 417-456.

147. Grantham, W. D., and Smith, P. M., Development of SCR Aircraft Takeoff and Landing Procedures for Community Noise Abatement and Their Impact on Flight Safety, *Supersonic Cruise Research '79*, NASA CP-2108, Part 1, 1980, pp. 299-333.

148. Carlson, H. W., and Mack, R. J., Estimation of Leading-Edge Thrust for Supersonic Wings of Arbitrary Planform, NASA TP-1270, 1978.

149. Manro, M. E., Bobbitt, P. J., and Kulfan, R. M., The Prediction of Pressure Distributions on an Arrow-Wing Configuration Including the Effect of Camber, Twist, and a Wing Fin, *Supersonic Cruise Research '79*, NASA CP-2108, Part 1, 1980, p. 72.

150. Robins, A. W., Carlson, H. W., and Mack, R. J., Supersonic Wings with Significant Leading-Edge Thrust at Cruise, *Supersonic Cruise Research '79*, NASA CP-2108, Part 1, 1980, pp. 229-246.

151. Roensch, R. L., and Page, G. S., Analytic Development of an Improved Supersonic Cruise Aircraft Based on Wind Tunnel Data, *Supersonic Cruise Research '79*, NASA CP-2108, Part 1, 1980, pp. 205-227.

152. Bower, R. E., Introductory Remarks, *Supersonic Cruise Research '79*, NASA CP-2108, Part 1, 1980, pp. 9-12.

153. Coe, P. L., Jr., Thomas, J. L., Huffman, J. K., Weston, R. P., Schoonover, W. E., Jr., and Gentry, G. L., Jr., Overview of the Langley Subsonic Research Effort on SCR Configurations, *Supersonic Cruise Research '79*, NASA CP-2108, Part 1, 1980, pp. 13-33.

154. Coe, P. L., Jr., Smith, P. M., and Parlett, L. P., Low-Speed Wind Tunnel Investigation of an Advanced Supersonic Cruise Arrow-Wing Configuration, NASA TM-74043, 1977.

155. Coe, P. L., Jr., Huffman, J. K., and Fenbert, J. W., Leading-Edge Deflection Optimization for a Highly Swept Arrow Wing Configuration, NASA TP-1777, 1980.

156. Sigalla, A., *op. cit.*, p. 831.

157. Heldenfels, R. R., Introductory Remarks, *Supersonic Cruise Research '79*, NASA CP-2108, Part 2, 1980, pp. 553-561.

158. Morino, L., and Tseng, K., Steady, Oscillatory and Unsteady, Subsonic and Supersonic Aerodynamics (SOUSSA) for Complex Aircraft Configurations, AGARD CP-227, 1978, pp. 3-1-3-14.

159. Stroud, W. J., and Sobieszczanski, J., Advanced Structural Sizing Methodology, *Proc. NASA CTOL Transport Technology Conference*, Hampton, Virginia, Feb. 28-March 3, 1978, NASA CP-2036, Part I, pp. 311-330.

160. Heldenfels, R. R., *op. cit.*, p. 553.

161. Haskins, J. F., Kerr, J. R., and Stein, B. A., Time-Temperature-Stress Capabilities of Composites for Supersonic Cruise Aircraft Applications, *Proc. SCAR Conference*, NASA CP-001, Part 2, 1976, pp. 799-828.

162. Fischler, J. E., Opportunities for Structural Improvements for an Advanced Supersonic Transport Vehicle, *Supersonic Cruise Research '79*, NASA CP-2108, Part 2, 1980, pp. 589-616.

163. Ascani, L. A., and Pulley, J. K., New Advancements in Titanium Technology and Their Cost and Weight Benefits, *Proc. SCAR Conference*, NASA CP-001, Part 2, 1976, pp. 757-782.

164. Driver, C., *op. cit.*, p. 2.

165. *Ibid.*, p. 2.

166. Bales, T. T., Hoffman, E. L., Payne, L., and Reardon, L. F., Fabrication Development and Evaluation of Advanced Titanium and Composite Structural Panels, NASA TP-1616, 1980.

167. Heldenfels, R. R., *op. cit.*, p. 555.

168. *Ibid.*, p. 555.

169. Sigalla, A., *op. cit.*, p. 824.

REFERENCES

170. Stewart, W. L., Introduction, *Proc. SCAR Conference*, NASA CP-001, Part 1, 1976, p. 338.
171. Sigalla, A., *op. cit.*, p. 823.
172. Driver, C., *op. cit.*, p. 2.
173. *Ibid.*, p. 3.
174. FitzSimmons, R. D., McKinnon, R. A., and Johnson, E. S., Flight and Wind Tunnel Test Results of a Mechanical Jet Noise Suppressor Nozzle, *AIAA Paper 80-0165*, 1980.
175. Wright, B. R., Sedgwick, T. A., and Urie, D. M., *op. cit.*, pp. 939-984.
176. Neumann, F. D., Toward a Second Generation Fuel Efficient Supersonic Cruise Aircraft: Design Characteristics and Feasibility, *Proc. SCAR Conference*, NASA CP-001, Part 2, 1976, pp. 849-866.
177. Sigalla, A., *op. cit.*, p. 831.
178. Neumann, F. D., and Whitten, J. W., A Family of Supersonic Airplanes— Technical and Economic Feasibility, *Supersonic Cruise Research '79, NASA CP*-2108, Part 2, 1980, pp. 833-854.
179. Driver, C., *op. cit.*, p. 4.
180. *Ibid.*, p. 4.
181. Doggett, R. V., Jr., and Townsend, J. L., Flutter Suppression by Active Control and Its Benefits, *Proc. SCAR Conference*, NASA CP-001, Part 1, 1976, pp. 303-333.
182. Goetz, R. C., Loads Technology for Supersonic Cruise Aircraft, *Proc. SCAR Conference*, NASA CP-001, Part 2, 1976, pp. 692-693.
183. Runyan, L. J., Middleton, W. D., and Paulson, J. A., Wind Tunnel Test Results of a New Leading Edge Flap Design for Highly Swept Wings—A Vortex Flap, *Supersonic Cruise Research '79*, NASA CP-2108, Part 1, 1980, pp. 131-147.
184. Driver, C., and Maglieri, D. J., *op. cit.*
185. Kelly, R., Supersonic Cruise Vehicle Research/Business Jet, *Supersonic Cruise Research '79*, NASA CP-2108, Part 2, 1980, pp. 935-949.
186. Driver, C., *op. cit.*, p. 4.
187. *Ibid.*, p. 5.

Selected Additional Reading*

AERODYNAMICS

Brown, C. E., and McLean, F. E., The Problem of Obtaining High Lift-Drag Ratio at Supersonic Speeds, *Jour. Aero. Sci.*, Vol. 26, No. 5, May 1959.

Brown C. E., McLean, F. E., and Klunker, E. B., Theoretical and Experimental Studies of Cambered and Twisted Wings Optimized for Flight at Supersonic Speeds, *Proc. of Second International Congress of the Aeronautical Sciences*, Vol. 3, Sept. 1960, pp. 415-431.

Carlson, H. W., A Modification to Linearized Theory for Prediction of Pressure Loadings on Lifting Surfaces at High Supersonic Mach Numbers and Large Angles of Attack, NASA TP-1406, 1979.

Carlson, H. W., and Middleton, W. D., A Numerical Method for the Design of Camber Surfaces of Supersonic Wings with Arbitrary Planform, NASA TN D-2341, 1964.

Dollyhigh, S. M., Theoretical Evaluation of High Speed Aerodynamics for Arrow Wing Configurations, NASA TP-1358, 1979.

Jones, R. T., Wing Plan Forms for High-Speed Flight, NACA TN-1033, March 1946.

Kulfan, R. M., and Sigalla, A., Real Flow Limitations in Supersonic Airplane Design, *AIAA Paper 78-147*, Jan. 1978.

McLemore, H. C., and Parlett, L. P., Low Speed Wind Tunnel Tests of a 1/10 Scale Model of a Blended-Arrow Supersonic Cruise Aircraft, NASA TN D-8410, 1977.

---

*An essentially complete listing of formal reports, contractor reports, articles, meeting papers, and company reports related to and sponsored by the NASA SCR and VCE programs can be found in NASA RP-1003 (1977) and NASA RP-1063 (1980), by Sherwood Hoffman.

SUPERSONIC CRUISE TECHNOLOGY

Sears, W. R., Aerodynamics, Noise, and the Sonic Boom, *AIAA Jour.*, Vol. 7, No. 4, April 1969.

Shrout, B. L., and Fournier, R. H., Aerodynamic Characteristics of a Supersonic Cruise Airplane Configuration at Mach Numbers of 2.30, 2.96, and 3.30, NASA TM-78792, 1979.

ENVIRONMENT (NOISE, SONIC BOOM, AND POLLUTION)

Carlson, H. W., Barger, R. L., and Mack, R. J., Application of Sonic-Boom Minimization Concepts in Supersonic Transport Design, NASA TN D-7218, June 1973.

Diehl, L. A., Reck, G. M., Marek, C. J., and Szaniszlo, A. J., Stratospheric Cruise Emission Reduction Program, NASA CP-2021, 1977, pp. 357-391.

Goodykoontz, J., Effect of a Semi-Annular Thermal Acoustic Shield on Jet Exhaust Noise, NASA TM-81615, 1980.

Goodykoontz, J., and Von Glahn, U., Noise Suppression Due to Annulus Shaping of an Inverted-Velocity-Profile Coaxial Nozzle, NASA TM-81460, 1980.

Knott, P. R., Brausch, J. F., Bhutiani, P. K., Majjigi, R. K., and Doyle, V. L., VCE Early Acoustic Test Results of General Electric's High Radius Ratio Coannular Plug Nozzle, *Supersonic Cruise Research '79*, NASA CP-2108, Part 1, 1980, pp. 417-452.

Maglieri, D. J., Carlson, H. W., and Hubbard, H. H., Status of Knowledge on Sonic Booms, NASA TM-80113, 1979.

McLean, F. E., Carlson, H. W., and Hunton, L. W., Sonic Boom Characteristics of Proposed Supersonic and Hypersonic Airplanes, NASA TN D-3587, Sept. 1966.

NASA Conferences on Sonic Boom Research, NASA SP-147, April 12, 1967; NASA SP-180, May 9-10, 1968; and NASA SP-255, Oct. 29-30, 1971.

Poppoff, I. G., Whitten, R. C., Turco, R. P., and Capone, L. A., An Assessment of the Effect of Supersonic Aircraft Operations on the Stratospheric Ozone Content, NASA RP-1026, 1978.

Proc. of Sonic Boom Symposium, *Jour. Acoust. Soc. America*, Vol. 39, No. 5, Part 2, May 1966.

Proc. of Sonic Boom Symposium, *Jour. Acoust. Soc. America*, Vol. 51, No. 2, Part 3, Feb. 1972.

Turco, R. P., Toon, O. B., Pollack, J. B., Whitten, R. C., Poppoff, I. G., and Hamill, P., Stratospheric Aerosol Modification by Supersonic Transport and Space Shuttle Operations—Climate Implications, *Jour. Appl. Meteorol.*, Vol. 19, No. 1, Jan. 1980, pp. 78-89.

SELECTED ADDITIONAL READING

PROPULSION

Calder, P. H., and Gupta, P. C., Engine Options for Supersonic Cruise Aircraft, *AIAA Paper 78-1054*, July 1978.

Cole, G. L., Atmospheric Effects on Inlets for Supersonic Cruise Aircraft, NASA TM X-73647, 1977.

Fishbach, L. H., Stitt, L. E., Stone, J. R., and Whitlow, J. B., Jr., NASA Research in Supersonic Propulsion: A Decade of Progress, NASA TM-82862, 1982.

Franciscus, L. C., The Turbine Bypass Engine — A New Supersonic Cruise Research Propulsion Concept, *AIAA Paper 81-1596*, 1981.

Howlett, R. A., and Smith, M. G., Jr., Advanced Supersonic Transport Propulsion Systems, *SAE Paper 771010*, Nov. 1977.

Johnson, H. W., and Conrad, E. W., NASA Engine Systems Technology Programs: An Overview, *AIAA Paper 79-928*, July 1978.

Krebs, J. N., and Allan, R. D., Supersonic Propulsion — 1970 to 1977, *AIAA Paper 77-832*, July 1977.

Westmoreland, J. S., and Packman, A. B., A Successful Step Toward an Advanced Supersonic Transport Engine — Acoustic and Emissions Results from the Pratt and Whitney Aircraft Variable Cycle Engine Program, *AIAA Paper 81-1593*, July 1981.

STABILITY AND CONTROL

Averett, B. T., Simulator Investigation of Arrow-Wing Low-Speed Handling Qualities, *Supersonic Cruise Research '79*, NASA CP-2108, Part 1, 1980, pp. 285-298.

Berry, D. T., A Summary of YF-12 Handling Qualities, NASA CP-2054, Vol. II, 1978, pp. 31-57.

Chalk, Charles R., Flying Qualities Design Criteria Applicable to Supersonic Cruise Aircraft, *Supersonic Cruise Research '79*, CP-2108, Part 1, 1980, pp. 251-267.

Gilyard, G. B., and Burken, J. J., Development and Flight Test Results of an Autothrottle Control System at Mach 3 Cruise, NASA TP-1621, 1980.

McGehee, J. R., and Dreher, R. C., Experimental Investigation of Active Loads Control for Aircraft Landing Gear, NASA TP-2042, Aug. 1982.

Rezek, T. W., Pilot Workload Measurement and Experience on Supersonic Cruise Aircraft, NASA CP-2054, Vol. I, 1978.

185

SUPERSONIC CRUISE TECHNOLOGY

STRUCTURES AND MATERIALS

Adelman, H. M., Sawyer, P. L., and Shore, C. P., Optimum Design of Structures at Elevated Temperatures, *AIAA Jour.*, Vol. 17, No. 6, June 1979, pp. 622-629.

Bales, T. T., SPF/DB Titanium Technology, NASA CP-2160, 1980.

Boeing Commercial Airplane Co., Study of Advanced Composite Structural Design Concepts for an Arrow Wing Supersonic Cruise Configuration, NASA CR-145192, 1978.

Del Mundo, A. R., McQuilkin, F. T., and Rivas, R. R., SPF/DB Primary Structure for Supersonic Aircraft (T-38 Horizontal Stabilizer), NASA CR-163114, 1981.

Greene, W. H., and Sobieski, J. S., Minimum Mass Sizing of a Large Low-Aspect Ratio Airframe for Flutter-Free Performance, NASA TM-81818, 1980.

Guess, M. K., Kaneko, R. S., and Wald, G. G., Advanced Materials and Fabrication Processes for Supersonic Cruise Aircraft, *Supersonic Cruise Research '79*, NASA CP-2108, Part 2, 1980, pp. 687-712.

Haftka, R. T., and Shore, C. P., Approximate Methods for Combined Thermal/ Structural Design, NASA TP-1428, 1979.

Payne, L., Fabrication and Evaluation of Advanced Titanium Structural Panels for Supersonic Cruise Aircraft, NASA CR-2744, 1977.

SYSTEMS INTEGRATION/CONCEPTS

Brown, R., Integration of a Variable Cycle Engine Concept in a Supersonic Cruise Aircraft, *AIAA Paper 78-1049*, July 1978.

Chacksfield, J. E., Arrow Wing — Its Potentialities and Drawbacks with Regard to In-Flight Aerodynamic Research, *Aircraft Eng.*, Vol. 49, No. 8, Aug. 1977, pp. 4-8.

Driver, C., Advanced Supersonic Technology and Its Implications for the Future, *AIAA Paper 79-0694*, March 1979.

FitzSimmons, R. D., The Advanced Supersonic Transport: What It Is and How It Compares, *Acta Astronaut.*, Vol. 4, No. 1-2, Jan.-Feb. 1977, pp. 131-143.

FitzSimmons, R. D., Technology Readiness for an SST, *AIAA Paper 78-356*, Feb. 1978.

FitzSimmons, R. D., Rowe, W. T., and Johnson, E. S., Advanced Supersonic Transport Engine Integration Studies for Near-Term Technology Readiness Date, *AIAA Paper 78-1052*, July 1978.

Goodmanson, L. T., and Sigalla, The Next SST — What Will It Be?, *AIAA Paper 77-797*, July 1977.

Horwitch, M., Clipped Wings — The American SST Conflict, The MIT Press, Cambridge, Mass., 1982.

Maglieri, D. J., and Dollyhigh, S. M., We Have Just Begun to Create Efficient Transport Aircraft, *Astronaut. Aeron.*, Feb. 1982, pp. 26-38.

SELECTED ADDITIONAL READING

Radkey, R. L., Welge, H. R., and Roensch, R. L., Aerodynamic Design of a Mach 2.2 Supersonic Cruise Aircraft, *Jour. Aircraft*, Vol. 15, No. 6, June 1978, pp. 351–357.

Rowe, W. T., Technology Development Status at McDonnell Douglas, *Supersonic Cruise Research '79*, CP-2108, Part 2, 1980, pp. 873–888.

About the Author

F. Edward McLean went to work at NACA as an aerospace technologist in 1948, only 8 months after Charles E. Yeager's historic supersonic flight. McLean was involved in supersonic research from that time until his retirement in 1978. His major contributions were in the areas of supersonic airplane design and sonic boom. He was a technical advisor in the Federal Aviation Administration's SST evaluations and was NASA representative to the U.S. Air Force during the development of the F-15 fighter aircraft. McLean was an original member of the SCR Program Office at the NASA Langley Research Center, and was head of the office from 1974 until his retirement. Among his many honors were the NASA Special Service Award for Exceptional Service in 1967 for his application of "near-field" technology to the reduction of sonic boom and the NASA Exceptional Service Medal in 1978 for his conduct of the NASA SCR program.